Articulating Difference

Articulating Difference

Sex and Language in the German Nineteenth Century

SOPHIE SALVO

The University of Chicago Press

Chicago and London

The University of Chicago Press, Chicago 60637
The University of Chicago Press, Ltd., London
© 2024 by The University of Chicago
Published 2024
Printed in the United States of America

33 32 31 30 29 28 27 26 25 24   1 2 3 4 5

ISBN-13: 978-0-226-82770-4 (cloth)
ISBN-13: 978-0-226-82772-8 (paper)
ISBN-13: 978-0-226-82771-1 (e-book)
DOI: https://doi.org/10.7208/chicago/9780226827711.001.0001

Library of Congress Cataloging-in-Publication Data

Names: Salvo, Sophie, author.
Title: Articulating difference : sex and language in the German nineteenth century /
    Sophie Salvo.
Description: Chicago : The University of Chicago Press, 2024. | Includes
    bibliographical references and index.
Identifiers: LCCN 2024011347 | ISBN 9780226827704 (cloth) | ISBN 9780226827728
    (paperback) | ISBN 9780226827711 (ebook)
Subjects: LCSH: German language—Sex differences—History—19th century. |
    Language and sex—Germany—History—19th century. | Women—Germany—
    Language—History—19th century. | Sexism in language—Germany—History—
    19th century. | Language and languages—Sex differences—History—19th century. |
    Language and languages—Sex differences—Philosophy. | Language and
    languages—Philosophy—History—19th century. | Grammar, Comparative and
    general—Gender—History—19th century. | Linguistics—Germany—History—
    19th century.
Classification: LCC PF3105 .S28 2024 | DDC 435—dc23/eng/20240318
LC record available at https://lccn.loc.gov/2024011347

♾ This paper meets the requirements of ANSI/NISO Z39.48-1992 (Permanence of Paper).

Contents

Introduction

The relationship between gender and language is a preoccupation of our time, yet as a subject of historical inquiry, it has mainly been studied obliquely. There is much that we know, for example, about the histories of pronouns, about the etymologies of words related to sex, gender, and sexuality, about historical definitions of *masculine and feminine*, and about the gendered histories of oratory, language pedagogy, and lexicography, not to mention the myriad of ways that women's writing has been denigrated historically.[1] But when it comes to the concepts that underwrite these phenomena—in other words, how sex, gender, and language have been linked at a theoretical level—scholars have rarely considered the topic directly. *Articulating Difference* argues that in order to understand something like the historical disparagement of women's writing or the success of the generic masculine pronoun, we must uncover the conceptualization of language that licensed it. That is the aim of this book: to investigate the histories of fields that take language itself as their focus, specifically philology, the philosophy of language, and linguistics, and examine how they theorized their subject in connection to gender. By laying bare the constitutive role of gender in a distinct era of linguistic thought, it seeks to reveal the inner workings of one pivotal chapter in the history of misogyny.

Nineteenth-century German language science (*Sprachwissenschaft*) is typically cited as a forerunner to the modern discipline of linguistics.[2] The period between the late 1700s and early 1900s witnessed extensive philosophical conjecture about language as well as new approaches to its analysis, including the development of the study of language into a "scientific" discipline. This discipline was concerned both with the development of individual languages and with the nature of language as such; it posited that a focus on the structural similarities of the world's tongues could reveal general truths about language, culture, and even humanity. When Sprachwissenschaft was being formed and practiced, however, ideas about the sexes often shaped how language was understood and analyzed. The notion—or, we might better say, fantasy—of sexual complementarity served as grounds for the production of knowledge about language and the humans who speak it. As scholars asserted the patriarchal origins of language, as they posited the existence of "women's languages" and valorized grammatical gender, they established a fundamental distinction between the masculine and the feminine as a necessary premise, relying on the assumed priority of the masculine to support their claims and delimit their methodologies. Wilhelm von Humboldt's theory of language, for instance, is structured around the binary active/passive—attributes that he derives from his characterizations of masculinity and femininity, respectively. He furthermore characterizes "women's language" in opposition to the precise scientific language of his own text. Contemporaries of Humboldt draw on ideas about the domestic economy of the heterosexual household to construct narratives about how language was first created and then dispersed. Philological studies of grammatical gender hinge on the assumption that femininity corresponds to weakness, passivity, and delicateness. *Articulating Difference* maintains that we should not, as has typically been the case, dismiss such gendered comments as expressions of personal sexual proclivities, as expendable metaphors for more serious philosophical problems, or as inevitable reflections of historical mores. Rather, we should recognize them as illustrative of a conceptual framework upon which thinking about language has depended, even what made this thinking possible. If we do not take the role of gender seriously, we not only misinterpret the history of language philosophy and linguistics, but we also risk reinscribing the same heteronormative, binary model of the sexes postulated by the historical texts. We risk assuming, in other words, that woman is man's natural complement, her passivity self-evident and unnecessary to interrogate.

That histories of linguistic thought have seldom paid attention to the assertions their primary texts make about sex and gender is no accident. It is a product of the structural continuities between nineteenth-century language science and its twentieth- and twenty-first-century historiography. While histories of language science and philosophy were produced in the nineteenth century,[3] the historiography of linguistics really took off in the later twentieth, spurred in part by Noam Chomsky's writings on Wilhelm von Humboldt in the 1960s.[4] Studies such as those by Hans Aarsleff and E. F. Konrad Koerner focused on differentiating historical schools of thought, on evaluating which ideas about language earlier thinkers got right, and, most importantly, on tracing the development of linguistics into the discipline we know today. When the journal *Historiographica Linguistica* was founded in 1974, it took as its charge a type of history writing that presents "our linguistic past as an integral part of the discipline itself and, at the same time, as an activity founded on well-defined principles which can rival those of 'normal science' (Kuhn) itself with regard to soundness of method and rigour of application."[5] In its emphasis on scientific method and rigor, and in its justification of its purview as a distinct (sub)discipline, the journal's description mirrors many self-descriptions of nineteenth-century language science, which I will discuss in more detail shortly. Even more important than its construction of "scientificity" is the narrative arc that it establishes. Whereas nineteenth-century authors are concerned with locating the origin(s) of language(s), linguistic historiography is concerned with locating the origin of linguistics; both suppose an ontology of progress organized around eminent figures and their inheritors, almost exclusively male. In the twenty-first century, this arc has been expanded to encompass new historical periods and national traditions and has faced new criticism—yet this has seldom meant a consideration of gender. While the past several decades have seen the publication of multivolume histories of linguistics, as well as similarly comprehensive studies of the history of language philosophy, these usually take the absence of women for granted.[6] Sarah Pourciau, who questions established accounts of the shift from diachrony to synchrony in linguistics, offers an important critique of such developmental narratives, yet rarely considers gender.[7]

Recently, scholars have begun to document the ways that women, while not welcomed into academic linguistics and language philosophy until the twentieth century, did contribute to the study of language from the margins.[8] This is important recuperative work, but it does not get at the issue in its entirety or its specificity. *That* women were excluded from linguistics and language philosophy is established fact, but *how* and *why* they were excluded is not simply explained by their long-standing political and educational disenfranchisement. There is, I would like to suggest, something particular about the period around 1800, when German linguistic thought was in its heyday: here, women's exclusion is embedded in the conceptualization of language itself. By aligning language with masculinity, linguistic texts justify, and indeed require, women's dehumanization on a theoretical level.

The historiography of linguistics cannot adequately account for its subject's history of exclusion if it reproduces structures of thought that, when they were forged in the nineteenth century, were done so through gendered frameworks and at woman's expense. Instead of outlining a history of ideas, therefore, which would classify the influences of various thinkers, or pinpoint the origin of methods and concepts, this book aims to uncover the theoretical and discursive preconditions of language science and philosophy in the nineteenth century. Thinking and writing about language, I aim to show, has often depended on an assertion of the sexes' inalienable difference.

Before going further, a note on terminology. Although *gender* is now common in English parlance, it was not employed outside of the grammatical context until the second half of the twentieth century.[9] In German, moreover, both sex and gender are expressed through the term *Geschlecht*, although the English-derived *Gender* is now also gaining prevalence. The German adjective *weiblich*, which means both *female* and *feminine*, poses a similar problem. Because of these chasms of history and of translation, it is difficult to standardize the terminology in a way that is both legible to twenty-first-century anglophone readers and faithful to the meaning of German historical texts. In this book, I favor clarity over exact systematicity and use the terms *sex* and *sexual differences* when referring to a text that makes claims about bodies, reproduction, and biology, *gender* when discussing a text's claims about femininity and masculinity in the abstract. This comes with the caveat, however, that for most of the texts read here, sex and gender cannot really be disambiguated. Indeed, as Judith Butler has argued, the idea that sex and gender can be completely disambiguated is erroneous in general. Although the common understanding of sex and gender is that sex is the "real" physical phenomenon, while gender is a social expression, sexual difference "is never simply a function of material differences which are not in some way both marked and formed by discursive practices." Sex, as Butler writes, is "not a simple fact of static condition of a body, but a process whereby regulatory norms materialize 'sex' and achieve this materialization through a forcible reiteration of those norms."[10] Of course, this is not the understanding of sex operative in nineteenth-century texts—they see sex as a naturally evident fact—but I include this discussion here to counter the contemporary impulse to starkly differentiate between the categories of gender and sex, whereby *sex* is made into something not constructed, but given.

Language Study and Its Disciplines

At the turn of the nineteenth century, Sprachwissenschaft was in its infancy. The very term for the discipline would be debated for decades. While August Schlegel and others wrote of *vergleichende Grammatik* (comparative grammar),[11] Wilhelm von Humboldt used *Sprachkunde* (the study of language), *Sprachwissenschaft* (language science), and *Sprachstudium* (language study) interchangeably,[12] August Schleicher argued for the Greek-derived *Glottik* (glottics), and other scholars favored the Latin-based *Linguistik* (linguistics).[13] This multiplicity of terminology reveals the extent to which the discipline, in its purview and its methods, was still in the process of being formed. Even the *Wissenschaft* of Sprachwissenschaft (for those who preferred this term) encompassed a range of approaches and practices. More capacious than our current use of *science* in English—which is typically restricted to the natural sciences—the German *Wissenschaft* refers to any systematic study or knowledge of a subject.

There is, furthermore, the issue of the relationship between Sprachwissenschaft and *Philologie* (philology). As "the discipline of making sense of texts"—to use Sheldon Pollock's definition—philology was a broad field that, depending on one's perspective, either included any study of language and language history or was restricted to the interpretation of historical works, particularly those of a national tradition.[14] Many scholars around 1800 considered Philologie to encompass what we would today call linguistics. Although the classicist Friedrich August Wolf "mentioned the terms *Linguistik* and *Sprachkunde* in his *Lectures on Altertumswissenschaft* [. . .] he took them to be interchangeable with philology," writes Ku-ming Kevin Chang.[15] Jacob Grimm at times used the name "neue Philologie" (new philology) to describe his linguistic inquiries.[16] By the second half of the nineteenth century, however, scholars were making a clear distinction between the two disciplines such that Schleicher would assert that "the philologist deals with history [. . .] the object of linguistics, on the other hand, is language."[17] While one could use the term *philology* to describe many of the texts examined in this book, I do so only when they explicitly identify themselves as such. Instead, I favor the terms *language science* and *linguistic thought* because these usefully capture the sense of a distinct discipline. Although they debated the terms and methods, the majority of the authors discussed in this book were committed to constructing and legitimating an autonomous discipline for the study of language, a discipline that they felt should mirror (if not exceed) the natural sciences. Investigations into historical languages were put in the service of comparative analysis, which could lead to revelations about the nature of human language and thus about the nature of humanity itself.

Around 1800, too, there was little formal or disciplinary distinction between what we would now differentiate as language philosophy versus linguistics. It was standard for texts to simultaneously provide information about specific languages, engaging in comparative analysis, and present abstract theories about language's nature and origin. For instance, Friedrich Schlegel's 1808 *Über die Sprache und Weisheit der Indier* (*On the Language and Wisdom of the Indians*), considered one of the founding texts of comparative, historical language study, includes detailed comparisons of the roots of multiple languages, which he refers to as *Beweis* (proof), as well as abstract claims about language, culture, and humanity.[18] The question of language's origin, which Schlegel also addressed in his 1808 text, was the subject of intense philosophical and scientific exploration in this period. As I discuss in chapter 1, eighteenth-century philosophers prized the form of the conjectural history—part theoretical treatise, part narrative—as a means of speculating about the beginnings of language. Even the more self-consciously philosophical investigations into the origin of language typically included discussions of grammatical structures. Johann Gottlieb Fichte's 1795 "Von der Sprachfähigkeit und dem Ursprung der Sprache" ("On the Linguistic Capacity and the Origin of Language"), for example, offers both a philosophically grounded account of language's origin and a discussion of verb tenses and noun cases.[19]

In the decades that followed, language science "came to find in the universities its natural habitat," as Anna Morpurgo Davies writes.[20] The first chair in general linguistics, held by Franz Bopp, was established at the University of Berlin in 1821. The discipline grew quickly, fueled by the general expansion of universities in Germany,[21] and by 1902, "all German Universities, ancient and new, had at least one official position in linguistics and often more than one."[22] As language science became institutionally recognized, its practitioners increasingly pushed to distinguish their discipline from other fields of study.[23] Conjectural theorizing was nonetheless commonplace in texts that define Sprachwissenschaft as "scientific." Emblematic here is the work of Jacob Grimm, who, although primarily concerned with comparing the sounds and structures of different languages, also engages in speculative explanations—about the origin of grammatical gender, for instance, as I explore in chapter 3. Similarly, Georg von der Gabelentz, whom I discuss in chapter 2, offers a spirited defense of Sprachwissenschaft in contradistinction to other fields of study, championing language science as engaged in the discovery of truths that other sciences cannot fathom; he also provides a theoretical meditation on the nature of language and its relation to human history.[24]

Toward the end of the nineteenth century, a new approach to language science appears on the scene: the Neogrammarians, a group of linguists based in Leipzig, who argue, among other things, that "sound change occurs according to mechanical laws" and that analogy "was as important a factor of language change in the past as it is in present-day languages."[25] The "Neogrammarian polemic," to use Sarah Pourciau's phrase, attempts to rid the discipline of speculative content and retrain its focus on data that can be observed and documented.[26] The programmatic preface to an 1878 Neogrammatical text criticizes the hypothetical reconstructions performed by earlier linguists and commands that "the comparative linguist must turn his gaze away from the *Ursprache* and towards the present."[27] This self-proclaimed "turn towards the present" does not constitute a total break from the linguistic thought that came before, as Pourciau argues. The move "merely reverses the prioritization of originary language without transcending it. The privileged linguistic object is now no longer the oldest but the 'youngest' and most alive."[28] Nevertheless, by the turn of the twentieth century, although many of the Neogrammarians' ideas had been criticized and dismissed,[29] their argument against hypothetical speculation had penetrated the field.

The outline sketched above is cursory and does not do justice to the many different schools and methodologies that developed in Germany throughout the nineteenth century. My intention is not to provide an exhaustive taxonomy of the history of linguistic thought, nor do I wish to enter debates about which author should *really* be considered the founder of modern German linguistics or philosophy of language.[30] That such an approach reproduces the same problematic structures of patrimony and inheritance found in the primary texts hardly needs to be stated. Instead, I aim to chart the rise and fall of a certain conceptualization of language—central to many linguistic studies of the period—and investigate its consequences for the imbrication of language and sexual complementarity.

As many critics have argued, chief among them Michel Foucault, the formation of the discipline of Sprachwissenschaft is tied to a shift in the conceptualization of language that occurs around 1800. Foucault famously describes this as the dawning of the modern episteme, where language was understood not as an imperfect tool that humans use to represent a preexisting reality, but rather as what makes the distinctly human experience of reality possible. In this formulation, language is "'rooted' not in the things perceived, but in the active subject. [. . .] a product of will and energy, rather than of the memory that duplicates representation." Foucault continues: language is "no longer linked to the knowing of things, but to men's freedom."[31] Thus would Johann Gottfried Herder, in 1771, call humans "linguistic creatures" and argue that human language was distinct from animal language from its very inception.[32] Or, as Jacob Grimm would put it in 1851, language is "human, it owes its origin and progress to our full freedom, it is our history, our heritage."[33]

The contours and claims of Foucault's theory have been rightfully debated.[34] Yet even if one rejects his specific terminology or framework, there are clearly significant changes to the way language is understood that are initiated in the late eighteenth century. In this period, the conceptualization of language is premised upon several related assertions, including: (1) that the difference between human and animal language is a difference in kind rather than type or level of development; (2) that there is a reciprocal relationship between the structures of language and the structures of human life; (3) that reason and language are interdependent; and (4) that language functions like an organism, driven by an internal principle of development, which can be studied scientifically, like the natural sciences study the phenomena of the natural world.

What happens to "woman" when language is conceptualized in this way, as fundamental to human nature, reason, and freedom? By scrutinizing the narratives, characters, and tropes that texts on language produce—that is, by reading philosophical and scientific texts from a literary-critical perspective—we will see "woman's" ambivalent relation to language, and, more broadly, to the category "human" in this period. Concomitant with the reconceptualization of language around 1800, I argue, is its alignment with masculinity.

There were, to be sure, many material reasons why women did not become linguists or philosophers between the late eighteenth and early twentieth centuries. First, there was the problem of educational and institutional access: women were officially barred from universities until the late nineteenth century, de facto until the early twentieth.[35] Then there was the problem of social norms and expectations, grounded in biologized notions of women's difference: certain fields, generally those that required conceptual analysis, were deemed contrary to women's nature.[36] The higher girls' schools established in the mid-nineteenth century typically restricted their curricula to "feminine" subjects like knitting and sewing.[37] Even upper-class women, who did not face the same incessant burdens of housework and child-rearing as did those in the lower classes, were nonetheless discouraged from pursuing theoretical and scientific disciplines, even as amateurs.[38] While these restrictions are important to register, they do not comprise my book's focus; this is a discursive rather than social history. It argues that the historical exclusion of women from language philosophy and science must be explained not only by educational prohibitions or restrictive mores but by the conceptualization of language itself—more precisely, by the fantasy of sexual differentiation upon which this conceptualization so often relied.

Gender and Language in the History of Knowledge

It is helpful to consider how German linguistic thought builds on ideas about sexual differentiation advanced by earlier intellectual traditions so that we may also identify how it diverges. Plato's *Timaeus* and Aristotle's *Generation of Animals*, for example, already establish the division of the active and the passive that would become relevant to later language philosophy. Medieval grammarians also use this binary in their discussions of grammatical gender.[39]

One can, moreover, find passing misogynist statements in a range of ancient and early modern texts on language. These include the claim entertained by Plato in the *Cratylus* that women don't create names as wisely as men, as well as Vico's reference to women's feeblemindedness (I return to these in chapter 2). Yet such descriptions do not comprise a systematic conceptual framework but rather form a kind of misogynist detritus that attaches to intellectual expression in an era where misogyny is common.

In many eighteenth- and nineteenth-century disciplines, however, ideas about sexual complementarity did play a systematic role in the production of knowledge. German Studies scholars have persuasively shown how masculinist thinking shaped a number of German discourses in this period. Stefani Engelstein, for instance, argues that the figure of the sibling became an epistemological tool for fields as diverse as philology, philosophy, literature, and biology. Wherever notions of the sibling were at stake, questions of sex and sexual exclusion were not far behind. Particularly during the revolutionary age of *fraternité*, "men could be re-created as citizens, equals, and affectively joined as brothers only because of the sequestration of sisters. The generation of sibling affect through women in order to bind together male agents as brothers structured politics and economics simultaneously."[40] Focusing on the significance of changing conceptions of sexual reproduction for biology, philosophy, and literature, Jocelyn Holland and Helmut Müller-Sievers each show not only the connections between scientific and literary discourses but also how they reinforce norms of masculinity and femininity.[41]

More broadly, since the 1980s, feminist scholars have emphasized the significance of an assumed sexual hierarchy for the formation of knowledge in an array of fields, including primatology (Donna Haraway), chromosome research (Sarah Richardson), and biology (Anne Fausto-Sterling).[42] Londa Schiebinger has investigated the histories of botany, anatomy, and other natural sciences to demonstrate how "scientists' claim to objectivity and value neutrality was the linchpin holding together a system that rendered women's exclusion from science invisible, and even made this exclusion appear fair and just."[43] Michèle Le Doeuff makes a similar argument regarding the history of philosophy, arguing that we must take seriously "stupid utterances made about women by people who, in principle, have no right to stupidity" and "bring to light what is at stake in those utterances."[44] Bonnie Smith and, more recently, Falko Schnicke have also discussed the systematic exclusion of women from the historical "sciences" in Germany, while Patricia M. Mazón shows the link between "academic citizenship and masculinity" in the establishment of the modern German research university.[45] Language philosophy and science, however, have received comparatively little attention in this regard.

When the relationship between notions of sex and language *has* been examined, it has primarily been in fields concerned with the present: sociolinguistics and feminist theory. Beginning in the 1970s, linguists such as Robin Lakoff researched the contexts and norms of women's speech, showing how American girls are socialized into speech patterns that limit agency and authority. In the following decades, sociolinguistic studies by Deborah Tannen, Sally McConnell-Ginet, and Deborah Cameron, among others, further analyzed and debated the interrelation of language and gender identity, concentrating on women's language use.[46] Sociolinguists began to study men's speech about a decade later.[47] Feminist psychoanalytic, social, and literary theories from the 1970s and 1980s shared many of the concerns of feminist sociolinguistics—such as charting the interrelation of language, power, and sexual difference—but offered more abstract accounts of women's language than empirical analysis. Hélène Cixous, for example, discusses the "multiplicity" and fragmentary nature of *écriture feminine* (feminine writing) in her now well-known essay, "Le Rire de la Méduse" ("The Laugh of the Medusa") from 1975. This text, which plays with descriptions of feminine language via tropes like breast milk and hysteria, was part of a larger, transnational reevaluation of gender, authorship, and canonicity taking place across feminist discourses in the 1970s and 1980s.[48]

I would like to dwell for a moment longer on Cixous's text because its reception illustrates something significant about the presentist bent of recent critical discussions of gender and language. In "Medusa," Cixous stages écriture feminine as a radical break from everything that has come before: "the future must no longer be determined by the past"; the text's reflections "[take] shape in an area just on the point of being discovered."[49] Therefore, she writes, "as there are no grounds for establishing a discourse, but rather an arid millennial ground to break, what I say has at least two sides and two aims: to break up, to destroy; and to foresee the unforeseeable, to project."[50] Scholarship on Cixous's écriture feminine—or on Luce Irigaray's somewhat analogous notion of *parler femme* (speaking woman), or on Julia Kristeva's idea of the semiotic[51]—has more or less taken this characterization at face value and considered their theories within a particular intellectual context. Works of psychoanalysis and poststructuralism provide the main interpretive framework.[52] Scholars have also read the theories of Cixous and Kristeva in connection with a specific literary canon of Modernist and women writers. These references make sense: they are the sources, often contemporary, explicitly mentioned by the primary texts themselves. But to read feminist theories of language only in the context of their own allusions obscures the ways in which they, too, participate in a practice, begun centuries prior, of using sexual difference as a premise for conceptualizing what language is and how it functions. That is not to say that these constructions are all the same, but that many of the ideas they rely upon—the existence of a distinct "women's language," the presumed masculinity of "regular" language, the reciprocal relationship between the structures of language and the structures of sex— have their own histories, histories that are interwoven with the formation of language science and philosophy in ways that are both unsettling and insufficiently acknowledged.

Tristram Wolff has written recently of "the literary history of linguistics," a phrase that I would adapt to encompass not only the "linguistic proposals" performed by literature but also the literariness of linguistics' proposals.[53] This is because, between the late eighteenth and the late nineteenth century, texts on language construct their philosophical and scientific systems by relying on the speculative, the *literary*, even as they deny their relation to the literary and define themselves against the ostensible fabrications of literature. As mentioned earlier, this period was famously preoccupied with the question of language's origin. Yet because this origin was considered an irrecoverable event in human history, it could only be addressed through conjecture. This posed a problem for philosophical and scientific works that professed to reveal only what was absolutely true, verified by empirical observation or fundamental laws. One way that fictional origins are transmogrified into scientific or philosophical truth is through their conformity to the universal "truth" of sexual complementarity.

We will see this occur throughout the book, so I will offer here just one illustrative example: the theory of grammatical gender put forth by Wilhelm Bleek, a German linguist working in South Africa. Bleek's general argument— that grammatical gender originates out of the personification of the world—is typical of nineteenth-century accounts. In his 1867 *Über den Ursprung der Sprache* (*On the Origin of Language*), he furthermore suggests that only languages with grammatical gender facilitate scientific achievement. Spinning a story about some of the first humans and their desire to make sense of the world around them, Bleek writes that they rendered inanimate objects male or female through the forms of grammatical gender. This personification incited "profound study," sharpening the human "power of observation," and paving the way for nothing less than a "knowledge of the final ground of all existence."[54] Notably, Bleek's own self-described "scientific" account rests upon neither empirical analysis nor inductive reasoning, but upon conjecture—a narrative about early humans and their drive to assimilate the alterity of the natural world. It is the assumed validity of the sexual hierarchy that allows Bleek's text to assimilate *its* speculative account of grammatical gender into the norms of scientific discourse, anchoring a tale that, in another context, might well be taken for fiction.

It is examples such as these that justify taking a literary-critical perspective on scientific and philosophical texts, unfolding their "poetics of knowledge," to use Jacques Rancière's term: "the set of literary procedures by which a discourse escapes literature, gives itself the status of a science, and signifies this status."[55] Or as Joseph Vogl similarly puts it in his call for a "poetology of knowledge," literary studies' object of analysis need not be restricted to a "type of text and institution that one might call 'literature.' Its subject is rather the literary, i.e., the relationship of language to itself and a fictive layer, in which what is seen and read, words and things are juxtaposed and interrupt the order of representation."[56]

In scholarship on the language sciences, critics have especially focused on the constructions of race in this "order or representation." Works by Ruth Römer, Maurice Olender, Geoffrey Harpham, Joseph Errington, Tuska Benes, and Markus Messling reassess the canonical texts of language science and philology, revealing the racialized—often racist—contexts out of which ideas about language were forged in Europe.[57] More recently, David Golumbia has made a case for recognizing both the "poisonous" and "emancipatory" racial legacies that persist in philology today,[58] while Siraj Ahmed argues that "philological approaches to languages, literatures, and religious traditions helped foster" colonial law in British India and "cleared the way for the political paradigm of emergency."[59] Like anthropology, nineteenth-century language disciplines depended on the colonial enterprise for much of their data and also advanced a binary of "primitive" versus "civilized" to substantiate their claims about language origin and development. Especially when the question of the Indo-European homeland was at stake, some thinkers "equated linguistic affinity with similarities of biological descent," in Benes's words.[60] In nineteenth-century linguistic thought, however, constructions of race are also tied to constructions of sex: one form of alterity is often made to substantiate the other. *Articulating Difference* brings out this complicated nexus of race, culture, and sex as it probes the theoretical masculinization of language that began around 1800.

Chapter 1, "Gendered Origins," investigates the disappearance of women from German philosophies of language in precisely this period. Whereas earlier theories typically posit a heterosexual pair at language's origin, beginning in the late eighteenth century, philosophical narratives illustrate their claims about the invention of language through male characters alone. Canonical thinkers such as Johann Gottfried Herder, Johann Gottlieb Fichte, and Wilhelm von Humboldt, moreover, define language analogously to how they define the masculine contra the feminine (e.g., characterized by reason, self-activity, creativity) and employ the model of the patriarchal, heterosexual family to explain how languages are created and develop. In this model, woman plays the role of passive recipient, facilitating and perpetuating the creations of men. As woman is excluded from philosophical narratives about the origin of language, she is made to bear an ambivalent relationship to the category "human." While woman is obviously human in the common sense of the term, she is not *emphatically* human, since she does not partake in the formation of language, that activity which most crucially distinguishes humans from animals.

Chapter 2, "'Women's Language' and the Language of Science," examines how the nineteenth-century human sciences make sense of women's deviation from the norms outlined in chapter 1—in other words, how they accommodate the fact that, in reality, women do not always play passive helpmate to the inventions of men but also form language themselves. I focus here on the term *women's language*, which first enters the European imagination through sixteenth- and seventeenth-century protoethnographic texts. These texts, which purport to document a "women's language" in the Caribbean and elsewhere, define it as a primitive phenomenon resulting from mass murder and mass rape: abducted and enslaved by the men of another tribe, women ostensibly continued to speak their native language and pass it down to their daughters, generation after generation. In nineteenth-century language science and anthropology, however, *women's language* is reconceptualized into a universal female phenomenon, as scholars conflate "primitive" women's languages with the presumed alterity of all women's speech. Scientific texts not only produce the fantasy of a universal women's language as a derivation and degradation of regular (male) language, but also use this fantasy as a way to define the contours of their own burgeoning disciplines. Women's language, characterized as euphemistic, unrigorous, and imprecise, is cast as a foil to the authoritative language of science.

Chapter 3, "The Sex of Language: Grammatical Gender," traces changing conceptualizations of grammatical gender between the seventeenth and nineteenth centuries to show how linguists not only presupposed ideas about the difference between the sexes but also used linguistic data to shore up this difference. While Enlightenment grammarians considered grammatical gender to have little semantic value, beginning in the late eighteenth century, scholars of language take gender to be a meaningful, and legible, category. Grammatical gender becomes a prized structure because it is assumed to provide insight into the culturally specific mindset of its creators. The fact that nouns have different genders in different languages, it is believed, reveals the uniqueness of each culture's worldview. That an era so invested in the idea of linguistic relativity among nations should take recourse to this notion to explain grammatical gender is perhaps not surprising. What is less well known is how texts on grammatical gender also used this structure to construct fantasies of masculinity and femininity, turning grammatical gender into a proxy for debating the centrality of sexual difference—and particularly the primacy of the masculine—to human identity. This comes to the fore in a debate over the origin and meaning of grammatical gender that occurred in the 1890s. By reconstructing the disagreement between Neogrammarians and "Romantic" linguists (the followers of Jacob Grimm), I show that the argument centers on whether "the human," as a category, is necessarily sexed.

Chapter 4, "Women Writing on Language," traces how gendered tropes, such as the masculinized subject of language invention, the imprecision of women's language, and language as patrimony, structured not just philosophical and scientific texts but also lives—as they were lived and as they are remembered. I examine linguistic studies by largely forgotten women scholars from the nineteenth and early twentieth centuries, including Clementine von Münchhausen, Carla Wenckebach, Carolina Michaëlis de Vasconcelos, and Elise Richter. The chapter focuses on their strategies of constructing textual authority in an era when language philosophy and science were coded masculine, such as creating ungendered subjects in their examples and diagrams. The chapter also meditates on the methodological challenges facing a study of women linguists from this period. Surveying the various approaches of feminist criticism to the historical problem of women's exclusion—such as writing counterfactual history, studying women's fictional texts and artworks for their theoretical or scientific content, and considering women's contributions to men's published works—I consider the limits and possibilities of these methods for the study of women's writings and their role in intellectual history.

The first four chapters analyze the masculinist fictions embedded in language philosophy and science, as well as women's responses to them. Chapter 5, "Modernism's Masculine Language Crisis," looks to where these fictions are invoked not as philosophical example or scientific evidence, but rather *as* fictions—in other words, when they are taken up by literary discourse. Certainly, representations of gendered language, and norms of female speech, appear in literature long before the turn of the twentieth century. One might think of Shakespeare's loquacious Rosalind, or Goethe's silent Ottilie, or Austen's frivolous Mrs. Bennet. In such cases, literary texts depict (and potentially dismantle) conventional ideas about how women speak. What happens in the period around 1900, however, is something more radical. In experimental works from the early twentieth century, German and Austrian writers use the idea of a distinctly "feminine language" to imagine the possibility of a non-arbitrary language that can access and express what regular language cannot. The chapter focuses on texts by Walter Benjamin, Robert Musil, and Hugo von Hofmannsthal to show how these authors' early experimentations with media and genre depend on the assertion of a utopian "feminine" language that borrows from and comments on notions of women's language formulated in the previous century. Here, the naive poetics of nineteenth-century linguistic discourse are replaced by a critical self-reflection.

I end this book in the early 1900s because this is where my subject ends: in the twentieth century, figurations of sex are no longer de rigueur for the scientific and philosophical conceptualization of language. That does not mean that they disappear altogether, however. In the Coda, I examine the afterlives of nineteenth-century concepts of language and sex in recent discussions of gendered language. This includes progressive language reforms, such as campaigns for gender-inclusive language, as well as the backlash against them. In Germany, for instance, the far-right populist party *Alternative für Deutschland* (Alternative for Germany, or AfD) has railed against what it calls "gegenderte Sprache" (genderised language), which it views as a threat to German identity. Such criticism is not exclusive to the AfD but has become part of mainstream conservative discourses across Europe and the Americas.

Articulating Difference is not a book about today's language debates, but it does aim to situate these debates within a broader historical perspective. To say that ideas about the sexes have historically played a role in the conceptualization of language is, most fundamentally, to say that *there is a history* of thinking about these two categories together—a history in which current gendered language reforms take part. Especially now, as conservatives argue against the "corruption" of languages through innovations such as gender-neutral and nonbinary pronouns, it is imperative to recognize that the modern study and conception of language has always been informed by "gender ideologies." It just has not been acknowledged as such because, for centuries, the ideology of sexual complementarity masqueraded as the natural and moral order.

Gendered Origins

Let's begin with a story. A father brings his son to a patch of flowers and points. "Rose," the father says. The son is silent. Later, the father issues a command: "Bring me a rose." The son returns to the patch but is unable to find that particular flower. Undeterred, he chooses another plant of a similar shape nearby. "Rose," he murmurs to himself. He uproots the flower and presents it to his father. The father is pleased. He acknowledges that the son has chosen correctly.

What will the father do with the rose? What function or ceremony awaits it? Moreover, why does he demand *this* flower, the quintessential symbol of romantic love,[1] from his progeny? Such questions remain unanswered in the story's source text, which mentions the rose anecdote twice but does not elaborate on its conclusion. This is, perhaps, because the story comes from neither a parable nor a fairy tale, but an essay by the philosopher Johann Gottlieb Fichte. More precisely, it resides in the middle section of "Von der Sprachfähigkeit und dem Ursprung der Sprache" ("On the Linguistic Capacity and the Origin of Language"), which Fichte published in 1795.[2] In this essay, Fichte aims to investigate not "that and how some language or other *might have* been invented," but rather "that and how language *must have* been invented."[3] This entails establishing the means in human nature "which would necessarily have to be available in order to realize the idea of language"[4] and how these means "would [. . .] have to be used in order for the goal to be achieved"[5]—a history of language outlined a priori. This approach leads Fichte to a host of conclusions, including that language originates in a social context and that it develops from "hieroglyphic" into arbitrary signs. As we shall see—indeed, as the rose example already intimates—Fichte's theory also requires that language be a masculine invention. For, at least according to his own conception of the sexes, the characters at the center of Fichte's narrative are necessarily male.

*

In the period when Fichte penned his essay, the nature and origin of language were not only popular topics of inquiry but also inextricable. To understand what language *is*—Human invention? Divine creation? Animal instinct?— required understanding where language came from and how it developed. Even philosophical studies of language that did not expressly take its origin as their focus typically included a discussion of how language first came into being.

The question of language's origin was often addressed through conjectural history. A distinctly modern form of reasoning that flourished in the eighteenth century, in particular,[6] conjectural histories allowed for speculation about the origins of constitutive aspects of human life, such as language, social order, and government, based on essential premises about human nature. Philosophers first established fundamental truths about human beings and then proposed and followed exemplary characters through a developmental narrative.

As a genre, conjectural history bears an uneasy relationship to both history and literature, two discursive modes with which it closely overlaps. On the one hand, such texts characteristically begin by renouncing facticity, making clear that they are interested in something other than a history of recorded events. "Let us begin by setting aside all the facts," writes Jean-Jacques Rousseau in the *Discourse on the Origin of Inequality* (1755), "for they have no bearing on this question. This kind of enquiry is not like the pursuit of historical truth, but depends solely on hypothetical and conditional reasoning."[7] Similarly, in his *Letters on the Aesthetic Education of Man* (1794–95), Friedrich Schiller concedes that the "state of brute nature" he describes "is not, I admit, to be found exactly as I have presented it here among any particular people or in any particular age. It is purely an idea; but an idea with which experience is, in certain particulars, in complete accord."[8] On the other hand, conjectural histories tend to blur the line between speculation and fact as their texts progress, switching, for instance, from the subjunctive or present tense into the past.[9]

As much as conjectural histories profess to be divorced from the realm of the factual, they also disavow the fictional. Denigrating literature is a common method of self-definition. Johann Gottfried Herder, for instance, explains in his *Treatise on the Origin of Language* that a writer must base all speculation on laws of nature; otherwise, he is simply composing a "philosophical novel."[10] In his "Conjectural Beginning of Human History," Immanuel Kant likewise disparages ungrounded hypotheses as the "draft for a novel"—or, more bitingly, "mere *fiction*."[11] Despite an avowed distaste for the literary, however, there is much that the genre of conjectural history also has in common with literary prose, including narrative exposition, plot development, and the construction of fictional characters. Considering conjectural histories through the lens of literary criticism not only highlights the interrelation of philosophical and literary discourses but also reveals the implications and limits of philosophical concepts. Fichte's essay, for example, shares a general form with the theories that precede his, but differs in its protagonists and its grammatical mood. Whereas earlier texts describe how human history *may* or *might* have unfolded, Fichte, as we know, insists that he narrates only what *must* have occurred.[12] And whereas Fichte, both implicitly and explicitly, represents men as language's originators, earlier theories typically construct a male-female dyad in order to explain language's invention.

As fictional byproducts of philosophical systems, the characters posited by these theories testify to the fantasies latent in the texts that construct them. It is not a coincidence that the shift from male and female to exclusively male characters mirrors the shift from the possible to the necessary, as exclusion is written into the text itself. The "must have" is produced as a category over and against the wrong kind of linguistic production of women, who—as Fichte has it—are incapable of innovation in any meaningful sense. For Fichte's text and others like it, the origin of language is a birth story in which femininity has no active role. Woman enters into the narrative only as a means for male development: like Fichte's rose, grammatically and symbolically feminine, an object of male exchange.

This chapter focuses on the disappearance of the heterosexual couple from theories of language origin and reads this disappearance as indicative of a broader shift around 1800: the alignment of language with masculinity and "masculine" faculties. My aim here is to show how this alignment takes place in the work of several prominent philosophers, including Herder, Fichte, and Wilhelm von Humboldt. Rather than focus on the entirety of their philosophical systems, I concentrate on the correspondences between their writings on language and their writings on the sexes. As the chapter details, their insistence on the masculinity of language leads not only to the dehumanization of women but also to a kind of bad reasoning, as they must account for this dehumanization in ways that are often at odds with the general tenets of their philosophy.

One Male, the Other Female

In order to understand how the characterization of language changes around 1800, we must first consider what precedes it. In the seventeenth and early eighteenth centuries, theories of the origin of language generally fall into two camps: those that claim that language has a divine provenance and those that argue that language is a human invention. Divine origin theories rely on biblical sources, extrapolating from Genesis about the speech of Adam and Eve. Such accounts align language with a general patriarchal authority—God, the ultimate father, endowed Adam, the father of humanity, with language. Some texts assert that Adam then passed this language onto Eve and their children, while others imply that Eve, too, received language directly from God.

Jean Frain du Tremblay's 1703 *Traité des langues* (*Treatise of Languages*) employs the latter explanation as it claims that "the origin of Tongues is owing to God alone."[13] The treatise imagines that Adam, the first man, had the ability to think—and thus a desire to speak—from the moment of his creation. Because humans are "made for Society," Adam "could have no Thoughts but what he was willing to communicate." God had thus endowed him with "certain Strokes of the Imagination proper to produce such Sounds, as were requisite to express them in such Manner."[14] However,

> because speaking alone is not sufficient to enable Man to Compose a *Tongue* (it being necessary likewise that Others should understand and comprehend what he says) God so ordain'd it, that at the same time, when the first man should speak, describing certain Things and expressing certain Ideas, his Wife and Children (when he should have them) should hear him speak, and at that time have in their Understandings the same Ideas, in their Brains the same Images, and in the Organs of their Voice the same Dispositions to pronounce the same Sounds; by which Means they should come to understand his Thoughts and to speak and answer him *Apropos*.[15]

The text places a noteworthy emphasis on *sameness*. Adam, his children, and his wife all have the "same Ideas," producing the "same Images" through the "same Sounds." Adam may gain the ability to speak before Eve—in addition to gaining "a full knowledge from God" of all the arts and sciences!—but she learns language not from a human father but a divine one.

William Warburton's *The Divine Legation of Moses Demonstrated* (1741), on the other hand, does not explain whether God gave the first man and woman language simultaneously. For Warburton, the biblical account of Adam's naming the animals should be understood as God *teaching* Adam language.[16] God "taught the first Man *Religion*; and can we think he would not, at the same Time, teach him *Language*?" Warburton asks. "Again, when God created *Man*, he made *Woman* for his Companion and Associate; but the only means of enjoying that Benefit was the Use of Speech. Can we believe that he would leave them to get out of the forlorn Condition of Brutality as they could?"[17] Whether Eve learned language from Adam or God directly, Warburton does not clarify.

Twenty-five years later, Jean Henri Samuel Formey also argues for a divine origin, claiming that nature can corroborate biblical authority. Contending that "there is no primitive language other than the one that the first man spoke because God had taught it to him,"[18] Formey proposes that a ruler (a prince, or barring that, a sovereign magistrate) should bring together a group of children who will be raised without language. This group, comprised of ten boys and ten girls,[19] will produce children, he insists, but never speech. For, Formey writes, "In order to speak, you must want to speak; how can you want to speak if you have no idea of speech?"[20] Without God's intervention, in other words, neither women nor men can create language for themselves.

Although arguments for the divine origin of language persist well into the nineteenth century,[21] beginning in the mid-1700s they are increasingly in the minority, eclipsed by theories that claim that language is a human invention.[22] Philosophers who endorse a human origin often rely on the idea of an original male-female couple but replace the biblical narrative with a narrative of their own creation. The notion of the human invention of language, in other words, begets the human-invented anecdote. Tracing the prevalence and function of the heterosexual pair in narratives of language creation from this period will allow us to more precisely reckon with its disappearance—and the disappearance of woman, in particular—in the period around 1800.

Already in John Locke's *An Essay Concerning Human Understanding* from 1689, a male-female dyad serves as the grounds for an exemplary narrative about language's development. Locke, who argued that language is an artificial, imperfect representation of ideas, is less interested in the origin of language per se than in how language might be improved. Nonetheless, he provides a conjectural account to illustrate the development of words of "mixed mode"—that is, words for complex ideas.

The anecdote takes Adam as its protagonist. Although Locke makes use of a biblical figure here, he is not arguing that language is a divine gift. Locke's Adam has no special wisdom; he is an ordinary human being who, as Hans Aarsleff writes, "is unprivileged in regard both to knowledge and naming."[23] Locke asks his reader to suppose the following: Adam observes Lamech, a descendant of Cain, acting melancholy and imagines that Lamech suspects that his wife, Adah, is unfaithful. In order to "discourse these thoughts" to Eve, and to ask Eve to "take care that Adah commit not folly," Adam coins two new words: *kinneah* and *niouph*, which stand for "suspicion, in a husband, of his wife's disloyalty to him" and "the act of committing disloyalty," respectively.[24] As it happens, however, Lamech was melancholy not because his wife was unfaithful, but because he had killed a man. Nevertheless, the terms *kinneah* and *niouph*, once introduced, grow into common use; they become the Hebrew words for jealousy and adultery. For Locke, the lack of any *original* act of jealousy or adultery illustrates how words are not tied to any essence.

In this narrative, the monogamous heterosexual unit plays an integral role. Locke recasts the biblical figure of Lamech—in Genesis he is a polygamist who "took two wives," Adah and Zillah—into a monogamist who has only Adah, the wife he "most ardently loved."[25] In Locke's version, the only promiscuity at issue in the Lamech passage turns out to be the promiscuity of language itself, as he reaffirms the fidelity of the married pair and explains instead the "disloyal" character of words. Not only is the marriage bond the topic of Adam's discourse, but it is also the context of its production. The premise behind Adam's coining new terms—namely, that Eve can understand the words as he creates them—requires her to possess the same faculties as he does, faculties that are no different from what humans possess today.[26] Could Locke equally have asked us to suppose that Eve, desiring to "discourse her thoughts" to Adam, invented a different word of "mixed mode"? Of course, Locke does not provide a direct answer to this in the *Essay*, but his writings elsewhere on the figures of Adam and Eve suggest that Eve's capacity for language is analogous to Adam's.

In the *First Treatise*, published the same year as the *Essay Concerning Human Understanding*, Locke argues against Eve's subordination to her husband. This comes as part of his refutation of another text, Sir Robert Filmer's *Patriarcha*,[27] which claimed that the divine right of kings derives from a paternal authority first held by Adam. In contrast, Locke contends that Adam's authority is much more limited and does not extend over his wife—at least not by divine sanction. In particular, Locke claims that God granted Eve *as well as* Adam dominion over the animals: when God "gave the world to mankind" he gave it to mankind "in common, and not to Adam in particular."[28] Furthermore, Locke maintains that even Eve's punishment after leaving Eden—"in sorrow thou shalt bring forth children, and thy desire shall be to thy husband, and he shall rule over thee"—did not performatively *grant* Adam authority over Eve, but merely "foret[old] what should be the woman's lot" according to the "laws of mankind and customs of nations."[29] As Ruth W. Grant points out, these customs are, for Locke, not rooted in natural or moral law, but rather in the artificial human construction of custom.[30] Locke's rendering of Adam and Eve here suggests that, except for differences in physical strength,[31] the capacities of men and women are at base the same, which includes their reason—and, by extension, their ability to form language.

In the century following Locke, this kind of conjectural anecdote becomes even more popular. Whether they argue that language develops out of instinctual needs or fervent passions, conjectural narratives from this period typically take recourse to the trope of the heterosexual pair. The survey of texts below, which spans the course of the eighteenth century as well as multiple national contexts, bespeaks the significance of the male-female dyad for thinking about language and its origins in this period.

The philosopher and satirist Bernard de Mandeville, for instance, relies on an exemplary male-female pair in his 1729 *Fable of the Bees*. Mandeville contends that language originated in the sounds of passion and sentiment, such as anguish, danger, grief, and compassion. These instinctive noises then developed slowly into sounds that refer to ideas and things. Mandeville accounts for this development by positing a "wild pair," male and female "savages," who, "when they have lived together for many Years," invent sounds to designate objects and communicate with one another. The couple teaches their words to their children, laying the groundwork for further invention, such that "every Generation would still improve upon."[32] Somewhat later in the century, the German scholar Johann David Michaelis would similarly tie the origin of language to the heterosexual couple. It is the love of the two sexes for each other, Michaelis writes in 1770, that "makes them eloquent and helps with new inventions."[33] Imagining the first couple in a state of nature—that is, with love but no marriage—Michaelis suggests that they "would now and then want to speak, and would try all kinds of signs, pantomimes and noises, sometimes to sweeten the passing of time, sometimes to summon a beloved object, to stop its leaving, or to overcome adversity by pleading and threats."[34]

Étienne Bonnot de Condillac, too, leaves the male and female roles in the creation of language explicit but undifferentiated. In his 1746 *Essai sur l'origine des connaissances humaines* (*Essay on the Origin of Human Knowledge*), Condillac asks us to suppose that, sometime after the great biblical Flood, "two children, one male, and the other female, wandered about in the deserts, before they understood the use of any sign."[35] According to Condillac, this youthful pair would first communicate their needs through instinctual cries and gestures, then become aware of how their cries were connected to "different motions of the body."[36] The children would eventually remember and reflect upon these sounds, which, Condillac maintains, would lead to their becoming habitual and to the development of new signs. In that Condillac insists on the children being "one male, the other female," he describes the creation of language as a mutual project undertaken by the heterosexual pair. Language develops out of what he calls their *commerce réciproque*, or reciprocal exchange.[37]

In Condillac, linguistic production mirrors sexual reproduction and vice versa. Over the course of the narrative, the pair matures into a young couple and Condillac asks us to suppose that they have a child—the natural course of events. The baby enters the world as his parents entered the desert, "pressed by wants which he could not without some difficulty make known."[38] Like language, which exists at this point only as a rude and deficient "language of action," the child is limited in his capacities and must undertake a slow development. The parallel between the couple's two creations highlights how, for Condillac, the creation of language is the result of carnal instincts. Language may be made *possible* by the human intellect, but what instigates its first invention is not deliberative reason, but rather the natural demands of the body—hunger, thirst, "the perception of a particular want."[39] Some years later, the Abbé Alexis Copineau draws a similar parallel between linguistic and sexual (re)production when he places male and female children at the center of his *Essai synthétique sur l'origine et la formation des langues* (Synthetic essay on the origin and formation of languages) from 1774. In this text, Copineau imagines what would happen if children were placed on a desert island, "a few European children of good constitution, different sexes, who had never heard human language."[40] He concludes that the children would develop language together out of a combination of cries, mimicry, and conventional signs.

Jean-Jacques Rousseau, on the other hand, locates the origin of language in human passions. Although Rousseau's "Essai sur l'origine des langues" ("Essay on the Origin of Languages"), published posthumously in 1781, does not propose a "first couple" like those of Mandeville or Condillac, it does posit a dialogical origin to language, arguing that humans desired to communicate "as soon as one man was recognized by another as a sentient, thinking being similar to himself."[41] Through instinct, then, humans developed "sensate signs for the expression of thought"—gesture and speech.[42] Rousseau names love and "stirring the heart" as two of the main passions that motivated humans to communicate,[43] thus implicating woman, as one half of the heterosexual pair, in this relational process. What's more, in his discussion of the origins of communication, Rousseau also provides a specifically *female* example: the daughter of Butades, an ancient Greek potter,[44] who traced the shadow of her lover, and thereby earned the title "the inventor of drawing."

Rousseau's pedagogical treatise *Émile* is notorious for its limiting view of women. "The quest for abstract and speculative truths, principles, and axioms in the sciences, for everything that tends to generalize ideas, is not within the competence of women," Rousseau writes there. "All their studies ought to be related to practice."[45] Yet this characterization of women's (in)competence does not exclude them from the origin of language as Rousseau imagines it.

First, because—according to Rousseau's account—humans do not originally develop language to communicate abstract truths and axioms but to express passions. Second, even if Rousseau's account of the origin of language *did* posit that the first words were invented via abstract speculation, women's incapacity for the theoretical is, in Rousseau's thought, primarily a consequence of women's socialization rather than their nature. In other words, significant differences between the sexes are "not the necessary *cause* of their social role but rather the intentional *result* of it," as Penny Weiss notes.[46] Or as Rousseau himself explains in *Émile*, the woman is weak and should be educated only in practical matters *because* she spends her time indoors—she "sees nothing outside the house."[47]

Adam Smith, in his *Considerations Concerning the Formation of Languages*, written in the 1750s, agrees with Condillac that humans developed language to communicate their immediate physical needs. In particular, Smith asks his reader to consider "two savages, who had never been taught to speak, but had been bred up remotely from the societies of men" who would "naturally begin to form that language by which they would endeavor to make their mutual wants intelligible to each other."[48] These "two savages" are ungendered in Smith's text. His examples of possible first words—*cave, tree, fountain*—correspond to shelter, hunger, and thirst, needs that are not restricted by sex. Indeed, the primary distinction upon which his text depends is not male versus female but "savage" versus "civilized." For Smith, gender—the social roles, and corresponding rights, of sexed beings—is not an a priori category but the result of a culture's level of "civilization." In his lectures on jurisprudence, Smith argues that women's liberty and authority change over time according to the society in which they live.[49]

The indifference to the difference of sex that we see in Smith takes an even more radicalized form in the work of the Scottish linguist Lord James Burnett Monboddo. Monboddo's *Of the Origin and Progress of Language* (1773) is now most famous for the connection it draws between humans and orangutans. Monboddo argues that language originated in the "natural and instinctive cries of the animal,"[50] which humans slowly refined and improved. Standing as evidence for this claim are observations about the sounds made by animals and by the people of "barbarous nations," who, according to Monboddo, use inarticulate cries to express "joy, grief, terror, surprise, and the like."[51] One of Monboddo's prime examples is that of the "savage girl": the so-called "wild girl of Champagne," a Native American child who was said to have spent a decade living in French forests, and whom Monboddo met and interviewed.[52] This girl, he writes, entertained me with several such cries belonging to her nation; and she told me, that, while she was traveling through the woods with the negro girl who had escaped the shipwreck with her, as they did not understand one another's language, they conversed together by signs and cries; and in that way they understood each other so well, that they made a shift to live upon what they could catch hunting together. These two methods of interpretation were undoubtedly the first used by men; and we have to suppose a great number of our species in the same situation as those two girls, carrying on some common business, and conversing together by signs and cries, and we have men just in a state proper for the invention of language.[53]

That two girls—and two non-European girls, at that—can symbolize the beginnings of the human species is worth dwelling upon. Of course, this beginning is implicated in a racialized model of human progress. The "savage" girls represent an earlier stage of human development, which Europe has apparently long surpassed.

By highlighting Monboddo's invocation of exemplary female figures, I do not mean to suggest that his text is necessarily championing women's rights. In the eighteenth-century world, the sexes were not equal socially, politically, or "naturally." As many historians have shown, women in this period were "subordinated to men, held to be unequal and treated unequally";[54] they were "presumptively unfit for most prestige-bearing and potentially powerful roles and opportunities, including the parish clergy, higher education, the military, many skilled crafts, village and town governance, and official state policy making—unless [they] happened to be an aristocrat or a queen."[55] This is not to say that nothing changed or was challenged during this period. There were moments of flux and debate over women's status—spurred in particular by Enlightenment feminist discourses[56] and by women's participation in the French Revolution—but this did not translate into radically improved conditions. Joan B. Landes argues that in the case of France, for example, "the eighteenth century marked a turning point for women in the construction of modern gender identity: public-private oppositions were being reinforced in ways that foreclosed women's earlier independence in the street, in the marketplace, and, for elite women, in the public spaces of the court and aristocratic household." Lieselotte Steinbrügge also observes that during the French Age of Enlightenment, women became increasingly associated with nature: "The accentuation of creatureliness, and thus of emotionality, over enlightened rationality predestined women to adopt [the] particular role" of a moral sex that was "efficacious through the private sphere."[57] Regarding German-speaking lands in the eighteenth century, Isabel Hull has similarly shown that "women's participation in the exercise of early civil society" was "strictly instrumental and relational," always considered in terms of male improvement.[58]

Although feminist voices across Europe objected to this asymmetrical treatment of men and women, the sexes nonetheless remained hierarchically organized in the cultural imagination. This hierarchy was premised on the idea that women are closer to nature and less capable of higher reasoning than men. Londa Schiebinger offers the telling example of Carl Linnaeus, who in 1758 coined the terms *mammalia* and *Homo sapiens* (man of reason): "Within Linnaean terminology, a female characteristic (the lactating mamma) ties humans to brutes, while a traditionally male characteristic (reason) marks our separatedness."[59] Like many scientists of this period, Linnaeus perpetuated the ancient idea that "woman—lacking male perfections of mind and body— resides nearer the beast than does man."[60]

Seen in this context, the appearance of women in the treatises cited above reveals more about their conception of language than their conception of the sexes. In a theory like Monboddo's (or Condillac's or Rousseau's, for that matter), the animal and the human, like the "primitive" and the "civilized," exist on a continuum. Because human communication develops out of animal communication—as "civilized" language develops out of "primitive" speech—there is little controversy to woman's role in its creation. Woman may be understood as less rational and less intelligent than man, but this does not preclude her from participating in the invention of language, since language does not originate out of reason or intelligence but, depending on which author you ask, out of need or passion.

Beginning in the late eighteenth century, however, contemporaneous with Monboddo's "savage girl" narrative, a new conception of language appears in German philosophy. Works by Herder, Fichte, and Wilhelm von Humboldt relate human and animal language not in terms of a continuum but a binary. In their theories, human language is distinct from animal language from its inception. "Even in its beginnings language is human throughout," Humboldt insists.[61] If woman is located somewhere between the animal and the human, her relation to language now becomes an ontological problem: woman and language are fundamentally misaligned. The changing cast of characters in theories of language makes this misalignment visible. Whereas earlier it was possible to illustrate the invention of language through different combinations of the sexes—a husband and wife, for example, male and female children, or even two girls—from this juncture onward, the creation of language will be postulated through male characters alone. At stake in the character construction of these theories is a question of representation: What constitutes the human, and who may stand in as example?

Herder and the Labor of Language

"The *woman*, in nature so much the weaker party, must she not accept law from the experienced, providing, language-forming man?"[62] So asks Johann Gottfried Herder's *Treatise on the Origin of Language*, which received the Berlin Academy prize in 1771. The question is rhetorical. "Indeed, is that properly even called law which is merely the gentle good deed of instruction?" That woman receives law and language from man is obvious. To know this, Herder writes, we need only think of the "economy of the nature of the human species."[63]

Herder's reference to the "economy" of the human race comes in the second half of the *Treatise*, where he considers how languages grow and spread: how the language of a family becomes the language of a tribe, then the language of a people and a nation. The first part of the *Treatise*, on the other hand, is dedicated to discussing the origin of language as such, including what makes language possible and how it relates to human nature. Here, Herder proposes a famous scene in which a human, encountering a sheep, invents language through the act of taking awareness.

This language-inventing character is not explicitly gendered. Recently, Herder scholars have even used feminine pronouns to refer to Herder's exemplary human, suggesting that we could understand his hypothetical first speaker as female.[64] While this practice allows for a certain kind of feminist intervention (asserting women's place in philosophical discourse), it forecloses another. Namely, it effaces the ambiguous status of woman in Herder's text and disallows an investigation into the relationship between his conception of language and his conception of the sexes. Whether Herder's human could be anything other than male—and whether language, and its entwined category, the human being, are necessarily aligned with masculinity—is a thornier issue than contemporary scholarship, and its rehabilitative pronouns, would tend to indicate. It is true that Herder does not exclude woman from intellectual life as summarily as will later figures such as Fichte and Humboldt. Nevertheless, if we read the *Treatise* not only as a theory of language but also as a theory of *labor*, we see that the act of creating language is men's work. The problem is not that women cannot or do not speak, but that their capacity for language is restricted to imitation. And imitation, Herder writes, is the province of animals rather than human beings.

FATHER'S LANGUAGE, MOTHER'S MILK

For all the attempts to recruit Herder as an exemplar of progressive gender politics,[65] this argument is complicated by a simple fact: Herder places language and patriarchy in a symbiotic relationship. It is the father who shapes language, and it is language that shapes the father's legacy. As Herder writes in the *Treatise*, language is the "characteristic word of the race, bond of the family, tool of instruction, hero song of the fathers' deeds, and the voice of these fathers from their graves."[66] In what follows, I examine Herder's constellation of language and paternal authority in part 2 of the *Treatise* and then consider the implications of this reading for our understanding of the theory of language invention that Herder puts forth in part 1. Throughout, I will refer to another of Herder's texts, his *Ideen zur Philosophie der Geschichte der Menschheit* (*Outlines of a Philosophy of the History of Man*),[67] which he published between 1784 and 1791, and which offers further insight into his conception of language, education, and the sexes.

Key to Herder's theory of language development is his concept of the *Familienfortbildung der Sprache*, or the family development of language. Although seemingly inclusive, *family* is in fact a metonym for the father and his authoritative structuring of the family. This identification of language with the father depends upon what Herder calls the "natural" division of labor within the familial unit. As Jennifer Fox observes, summarizing Herder's theory of parental responsibility, "the maternal province is to provide physical nourishment by the breast, the paternal role is to provide spiritual nourishment by instilling tradition."[68] Twice in the *Treatise*, Herder differentiates parental roles through the trope of the mother's breasts and the father's knees, knees that "come towards [the infant] to take him up as a son."[69] Or as Herder later puts it in the *Outlines of a Philosophy of the History of Mankind*, "the father now becomes the instructor of his son, as the mother had been his nurse."[70] It is not a coincidence that the father takes over the work of education as the son moves from infancy to childhood, as this is when the child learns to speak.

In Herder's work, fatherhood is not metaphorical; it is not a role that can be taken up by anyone. The decoupling of sex and gender, which some scholars have located in Herder's text "Liebe und Selbstheit" (Love and selfhood),[71] is not to be found in his writings on language and the family. Herder's conception of the Familienfortbildung der Sprache is rooted in biology and sex determines a parent's function. The mother's function is corporeal: gestation, lactation, attending to the physical needs of the child. This is the destiny of all women. A "sweet maternal affection, bestowed on woman by Nature," Herder writes in the *Outlines*, is "almost independent of cool reason, and far remote from the selfish desire of reward."[72] One consequence of this familial division of labor, I will suggest, is that while the mother may be a conduit for the perpetuation of language, only the father can be its inventor.

For Herder, the invention of language is a point in prehistory as well as an ongoing process. In both cases, the invention of language is linked to male figures. "We are all his sons," he writes, for example, referring to the "first father of the first tribe." "From him begin species, instruction, language. [. . .] we have all invented, formed, and deformed in his wake."[73] Herder again relies on a patriarchal cast of characters when he considers the nature of language today versus at its inception:

> The first father, the first needy inventors of language, who sacrificed the work of their souls on almost every word, who everywhere in the language still felt the warm sweat which it had cost their activity—what informant could they call upon? The whole language of their children was *a dialect of their own thoughts, a paean to their own ideals*, like the songs of *Ossian* for his father *Fingal*.[74]

The invention of language makes men sweat; it is intellectual labor. The fruits of this labor pass through the male line, rendering language a kind of patrimony. Benedict Anderson argues that for Herder, who writes that "every nation has its language," the "conception of nation-ness [is] linked to a private-property language"—each nation "owns" its language.[75] As we see above, it is the labor of the father that makes this eventual "national" ownership possible. It bears mentioning here that *father's sweat* is a term in German: *Vaterschweisz*, which the Grimms' dictionary defines as "sweat squeezed out of the father due to concern about the children, concern for the children themselves."[76] The dictionary's correlative entry for the mother, however, is not *Mutterschweisz* (mother's sweat) but *Mutterthränen* (mother's tears). Here, as in Herder, the physical labor of the mother—who surely expended some sweat of her own to bring her children into the world—is eclipsed by the cognitive labor of the father.

Several pages later in the *Treatise*, Herder again brings up the sweat of the father while criticizing Rousseau's theory of property, which claimed that private property is a social rather than natural phenomenon. Herder counters Rousseau here with typical rhetorical flourish: "Ask why does this flower belong to the bee that sucks on it? The bee will answer: Because nature made me for sucking!"

> And if now we ask the first human being, Who has given you the right to these plants?, then what can he answer but: Nature, which gives me *the taking of awareness*! I have come to know these plants with effort! With effort I have

taught my wife and my son to know them! We all live from them! I have more right to them than the bee that hums on them and the cattle that grazes them, for these have not had all the effort of coming to know and teaching to know! Thus every thought that I have designed on them is a seal of my property, and whoever drives me away from them takes away from me not only my life, if I do not find this means of subsistence again, but really also the *value of my lived years*, my sweat, my effort, my thoughts, my language. I have earned them for myself! And should not such a signature of the soul on something through coming to know, through characteristic mark, through language, constitute for the first among humanity more of a right of property than stamp on a coin?[77]

Cognition functions as a kind of ownership that is constituted in and passed down through language, which gives man even greater rights to the natural world than animals could claim. What Herder refers to here as the "seal of my property" he will later call the "seal of truth"—the distinctly human truth of coming to know sensuous objects by designating them "with the stamp of my *inner sense*." "Our whole life, then," Herder explains elsewhere, "is to a certain extent *poetics*: we do not see images but rather create them."[78] Hence the naturalness of property: man "owns" the plants he encounters because to know them is to create them. Language, of course, is the medium that makes such knowing and creating possible.[79] Crucially, the cognition that Herder describes above, the effort it demands and the value it produces, all originate in the father. "With effort I have taught my wife and son to know them!" exclaims Herder's imaginary patriarch. How do they receive this teaching? Comparing the case of the son to that of the wife will reveal woman's vexed relationship to the origin and development of language—and, correspondingly, to the "ideal of the individual who has realized himself in all respects"[80] around which Herder's theory revolves.

FORMATION, IMITATION

If the *Treatise* establishes language as patrimony, the *Outlines* explains the means by which this patrimony is perpetuated: "fatherly love," a virtue "that is best displayed by a manly education." In this way, the father "inures his son to his own mode of life: teaches him his arts, awakens in him the sense of fame, and in him loves himself, when he shall grow old, or be no more."[81] In short, fatherly love enables the reproduction of the individual and the tradition of his work. As the son learns from the father, he becomes the father himself.

Initiating the son into the father's way of life means initiating him into language, as language is the means by which reason is developed and the means by which man improves himself and his species. Language is both genetic and organic, passed down by fathers, but reshaped by sons, who transform what they receive "into their own nature."[82] This second step provides the distinctly human element. Otherwise, "how poor must the creature be, who has nothing of himself, but receives everything from imitation, instruction, and practice, from which he is moulded like wax!"[83] Indeed, Herder singles out the inability to move beyond imitation as a key distinction between humans and apes, who might otherwise rival humans in their abilities.[84]

People who speak a "learned" language, Herder writes—that is, those who speak a language that they do not themselves form—are bound to "wander, as if their reason were in a dream":

> They think with the reason of others, and are but imitatively wise: for is he, who employs the art of another, himself an artist? But he, in whose mind native thoughts arise, and form a body for themselves; he, who sees not with the eye alone, but with the understanding, and describes not with the tongue, but with the mind; he, who is so happy as to observe Nature in her creative laboratory, espy new marks of her operations, and turn them to some human purpose by implements of art; he is *authentically a man* [*ist der eigentliche Mensch*], and as such seldom appear, he is a god among men. He speaks, and thousands lisp his words: he creates, and others play with what he has produced: he was a man, and children perhaps come after him again for centuries.[85]

Is woman an *eigentlicher Mensch*? According to Herder, many men are also not "authentic humans" because they do not create language themselves but only employ "the art of another." But while men are inhibited by their personal limitations, or what Herder calls their individual powers,[86] women, as we shall see, are inhibited by the limitations of their sex.

As Herder conceives of it, "manly education" is not just a matter of individual development. The son's education is a "public, an eternal work": "It transmits to posterity all the excellencies and prejudices of the human species."[87] Herder's conflation of human education, in general, with "manly" education, in particular, leaves woman's development an open question. The exclusion of the female subject from the education of the human race, however, does not seem to pose a problem for Herder's text, since there are hardly any female subjects to educate. Curiously, Herder's fathers rarely produce daughters.[88] In contrast to the figure of the father, whose maturation the text traces from boy to man, wives and mothers are conjured fully formed. They spring forth from Herder's pen like impotent Athenas, bound to facilitate the formation of their male relations.

Woman participates in the development of language, and therefore of the human species, only insofar as she creates and nurtures a son who will

undergo the kind of "manly education" outlined above. Whereas the son relates to the father through a model of apprenticeship, in which the student
eventually supersedes the teacher, the wife relates to her husband through a
model of tutelage, in which he always retains authority. We see this unfold, for
instance, as Herder discusses the virtues of women in the *Outlines*. Although
he stresses that the status of women varies greatly around the world,[89] he nevertheless asserts several universal female qualities. These include women's
love of adornment and decoration, their cleanliness, and a "gentle endurance"
of their male counterparts.[90] "Happy is it," indeed,

> that Nature has endowed and adorned the female heart with an unspeakably
> affectionate and powerful sense of the personal worth of man. This enables
> her to bear also his severities: her mind willingly turns from them to the con-
> templation of whatever she considers as noble, great, valiant, and uncommon
> in him: with exalted feelings she participates in the manly deeds, the evening
> recital of which softens the fatigue of her toilsome day, and is proud, since she
> is destined to obedience, that she has such a husband to obey. Thus the love of
> the romantic in the female character is a benevolent gift of Nature.[91]

Woman's gentle obedience is most perfectly exemplified among so-called
"wretched" peoples, Herder clarifies, since the "abuses of civilization" tend
to corrupt the natural way of things.[92] Even with this caveat, however, the
description of woman that Herder offers here is a universal one. Obedience
is woman's nature. It is also natural to the marital relation that a woman's
experience of the world be mediated through her husband, exemplified in
this scene by the husband's evening recital of his "manly deeds." The wife
participates in this action—and the language that makes it possible—only
indirectly, as a silent recipient. Mutely she listens, feeling an "unspeakable"
affection. Once more, the fatherly chorus resounds: *With effort I have taught
my wife and son to know them!*

The marital relation itself is a naturally given phenomenon, according
to Herder. Once man reaches the age of maturity and "quits the house of his
father,"

> Nature drives man out to construct his own nest. And with whom does he
> construct it? With a creature as dissimilarly similar to himself, and whose pas-
> sions are as unlikely to come into collision with his, as is consistent with the
> end of their forming a union together. The nature of woman is different from
> that of the man: *she differs in her feelings, she differs in her actions.* Miserable
> he, who is rivaled by his wife, or excelled by her in manly virtues! She was
> destined to rule him by kindness and condescension alone [*nur durch nachge-
> bende Güte*], which render the apple of discord the apple of love.[93]

Whether we take the "apple of discord" to be a mythological or a biblical reference,[94] the meaning is clear. Nature requires woman to stay within her role as man's helpmate; she cannot enter into competition with her husband. Instead, she must "yield" (*nachgeben*)—saying, with Milton's Eve, "With that thy gentle hand / Seized mine, I yielded, and from that time see / How beauty is excelled by manly grace / And wisdom which alone is truly fair."[95] When Herder writes that man constructs his nest with woman, this is not a mutual project in the sense of the two beings doing the same work, not Condillac's commerce réciproque. On the contrary, in the construction of the household, as in the construction of language, woman follows man's lead and acquiesces to his instruction. Hence the query with which I began this section: Must not woman receive law from language-forming man? Whereas the son recasts and reforms what he learns from his father, the wife accepts and perpetuates it unchanged. Herder's characterization of women as essentially *receivers*—of language, tradition, education—is significant because it destines them to be perpetual imitators. And what is an imitator but an inauthentic human?

The notion that women receive rather than form language, which is suggested by Herder's theoretical work, is borne out in his more impressionistic writings on women. In 1770, in a letter to his then-fiancée Caroline Flachsland, Herder writes that, above all, he has "more disgust for a learned woman than for any creature in the world."[96] While Herder also looks unfavorably upon "learned" men, they do not inspire the same level of disdain. The "learned woman" is an impossibility. Several years earlier, as Birgit Nübel documents, Herder had written:

> A woman who is well educated but not learned will speak of things that are in her sphere with a fluency, unaffected certainty, and naive beauty that pleases her; but if a scholar comes who wants to weigh her words: she will be shy; if he wants philosophical explanations and determinations; she will stammer— stammer again, and finally repeat the same word [. . .] she is used to thinking about her world clearly but not logically, to speaking understandably and beautifully, but not learnedly and with careful precision.[97]

As Herder has it, the speech of a well-brought-up woman is naive and unaffected, flowing forth beautifully and quickly. Her language, in other words, is unmediated: she is a conduit that allows language to pass through and leaves no mark. This is why woman cannot be induced to provide philosophical explanations, and why, as Herder writes in different letter to his fiancée, "all things, all matters, all *sciences* are never for women."[98] Instead, Herder explains, women should be concerned only with that which makes them pleasing wives and good mothers.[99] The scholar therefore errs in asking the woman to perform a scholarly mode of speech, not only because she is incapable of thinking (and thus of speaking) in a logical and precise way, but also because this would require her to remold and reshape her language, and molding and shaping is not in her nature. This anecdote represents, in nuce, woman's function in the production of language: she repeats the language she has been given but cannot move beyond repetition.

Although Herder uses the phrases "brought up" and "used to" here, woman's relation to language is not the result of what he elsewhere calls "voluntary or societal forces"[100] but of an ontology determined by nature. Indeed, the figure of a learned woman is for Herder so grotesque, such an "abhorrence of nature," that to deal with her ultimately requires a complete change in register and a recourse to racialized wisdom. "A crowing hen and a learned woman are bad harbingers," Herder quips to Caroline, paraphrasing an "Arabic" saying. "Cut off both their necks!"[101] Behind the comedic lurks the threat of a violence necessary to ensure the stability of the natural order.

Herder reprises the hen analogy some years later in a review of the poetry of Sophie Mereau, whom he praises specifically because she "never goes beyond the limits of her sex."[102] Here, Herder writes that "a young man who imitates woman, whom he can never really impersonate, is despicable to women, just as men are repulsed by the hen that crows like a rooster."[103] This comparison suggests that the difference between the sexes is so significant that it supersedes even the difference between species. It also implies a link between women and animals that does not pertain to men (notice that only the woman is transformed within the analogy; the man retains his human identity).

As he asserts a model of heterosexual complementarity in the arts, Herder would appear to contradict the argument that I have so far been making about his work—namely, that women do not participate in the formation of language. "Since one sex cannot and should not work in the place or existence or way of the other," Herder writes,

> both must rather support each other, work hand in hand in the scope of the spirit, in the formation of the senses, principles, and morals: thus history sufficiently shows us that in Greece and Italy, in France and England, female hands also contributed to the altar of the Graces, that is, the female muse helped [*mitgeholfen*] with the formation and refinement of language, taste, morals, imagination, even the formation and refinement of practical principles themselves.[104]

Women have participated in the creation of all aspects of culture, Herder seems to suggest, from taste to morality to language. Yet if we look closely, it is difficult to actually isolate woman's role here. Though the passage is framed as a discussion of women's writing, in Herder's most consequential claims, women themselves are nowhere to be found. Instead, it is the female muse, the traditional inspiration for male artistic production, whose contributions he lauds, and disembodied female hands that pray at the altar of the feminine graces. By exchanging embodied sex for the abstraction of gender, human women for a divine femininity, Herder excises women from the historical narrative and complicates the very claim that he appears to advance. Moreover, even femininity, represented through the figure of the muse, does not itself *create*. The muse's function is to provide assistance—she is said to have *mitgeholfen*. In the "formation and refinement of language," she is no protagonist.

HUMANS, ANIMALS, AND WOMEN

In Herder's thought, both the original invention of language and its continuous formation are necessary expressions of human nature. Humans are "linguistic creatures";[105] the ability to create language is the "distinguishing trait" of the human being's essence.[106] As we have seen, the continuous formation of language is bound up with the patriarchal family structure. This holds true not only for the human species in its infancy, when it was made up of discrete clans, but also for humanity today. The "manly" education upon which the continuous formation of language depends is anchored in the nuclear family but also exceeds the individual family, becoming a model for cultural transmission at large: "Just as in all probability the human species constitutes a single progressive whole with a single origin in a single great household economy, likewise all languages, too, and with them the whole chain of civilization."[107]

When Herder proposes a scene to illustrate the original invention of language in part 1 of the *Treatise*, however, the family plays no role. Instead, he presents us with a solitary figure, the human being in isolation. Although this human is faced with an "ocean of sensations," he is able to separate one object, distinguish its "characteristic mark," and turn this into a grasped sign. Herder illustrates this event through a narrative that would live on as the text's most famous passage. "Let that lamb pass before his eye as an image," Herder proposes. Something happens to him "as to no other animal." Unlike the hungry wolf or the aroused ram,

> no instinct disturbs him, no sense tears him too close to the sheep or away from it; it stands there exactly as it expresses itself to his sense. White, soft, woolly—his soul, operating with awareness, seeks a characteristic mark—*the sheep bleats!*—his soul has found a characteristic mark. [. . .] and now the soul

recognizes it again! "Aha! You are the bleating one!" the soul feels inwardly. The soul has recognized it in a human way, for it recognizes and names it distinctly, that is, with a characteristic mark.[108]

This anecdote serves to highlight a fundamental distinction between humans and animals. There is such a thing as animal language, expressed in the immediate cries of pain, joy, and other passions. We retain these sounds, what Herder calls nature's "raw materials," in our interjections, for example.[109] But contrary to Condillac and Rousseau, Herder argues that such "animal" sounds are not the material out of which human language develops. Instead, human language originates in the kind of awareness exemplified in the passage above, an awareness that is the "primary force and activity of the human soul,"[110] which is made possible by the human being's lack of instinct. Interestingly, it is only in his discussion of animal sounds that Herder calls up explicitly female examples: women in mourning, their lamentations and "repeated Hallelujahs."[111] Animal sounds, Herder furthermore maintains, find a particular reception among children, the sick, the depressed, the lonely—and women.[112]

But what of the inventor of *human* language, that lamb-encountering individual? Herder never explicitly claims that women lack the basic capacity for reflection required to create language. Nevertheless, his characterization of familial labor, along with the term he uses to describe this "taking of awareness," suggest that the invention of language is—to borrow a phrase from Herder himself—"naturally and necessarily" male.[113] *Besonnenheit* is Herder's name for the condition that makes humans human and allows for the creation of language. The word is difficult to translate. In English versions of the *Treatise*, *Besonnenheit* is typically rendered as "awareness," although this does not quite capture the layers of meaning that have accreted over centuries. *Besonnenheit* is the German word for the Ancient Greek *sophrosyne*,[114] the virtue of self-control, or "self-knowledge and self-restraint" in Helen North's formulation.[115] In Herder, *Besonnenheit* is discussed as a condition rather than a virtue; it is "the state of mind of man in reflection."[116] Yet Herder, who was well versed in and even translated Ancient Greek literature,[117] would undoubtedly have been familiar with the original meaning of *sophrosyne*, which is referenced in texts from Homer to the tragic poets to Plato.

In the *Treatise*, Herder does not explain his use of *Besonnenheit*, writing only that he generally finds terminology unimportant:

> Let one name this whole disposition of the human being's forces however one wishes: understanding, reason, taking-awareness [*Besinnung*], etc. It is indifferent to me, as long as one does not assume these names to be separate forces

or mere higher levels of the animal forces. It is the *"whole organization of all human forces; the whole domestic economy of his sensuous and cognizing, of his cognizing and willing nature."*[118]

There are, nevertheless, good reasons for Herder's choice of the term, which reveal themselves in the following pages. Introducing Besonnenheit allows Herder to de-emphasize concepts like reason (*Verstand*) and reflection (*Reflexion*), which were employed by earlier philosophers such as Locke to advance theories of language with which he disagrees. As a state of being rather than a single faculty, *Besonnenheit* captures the idea that "reason is no compartmentalized, separately effective force but an orientation of all forces that is distinctive to [the human] species."[119]

Describing the invention of language as predicated on Besonnenheit also has another consequence: it further aligns the human, and his language, with something traditionally masculine. In the ancient world, sophrosyne was considered a female as well as male virtue, but it took a different form in women than in men. In men, sophrosyne manifested as "the harmonious product of intense passion under perfect control."[120] A *sophron* woman, on the other hand, was "dutiful, obedient, well-behaved."[121] This meant chaste as well as quiet.[122] It is clearly the male version of sophrosyne that Herder draws on in his conceptualization of Besonnenheit. For in Herder's anecdote about the exemplary first speaker, what is important is that man is not controlled by or obedient to anything: he is calm and free to reflect on the world around him, unmoved by instinct or physical need.

In Herder's naturalist philosophy, the anthropological reveals the ontological. In reference to human labor, we might say that effort reveals essence. Woman does a different kind of work than man—the work of mothering her son and the work of assisting her husband. In both of these activities, her labor is externally oriented, contingent on a male subject. At its most essential, woman's ontological condition is the condition of being in relation. Compare this to Herder's language-inventing human, who is complete within himself. His language neither comes from an external source (God), nor does it require an external interlocutor. On the contrary, the *Treatise* insists that, even if this figure never met another human, language still would have been invented. Language is dialogical, but this is an internal dialogue, a conversation of the self with the soul.[123] Although it is catalyzed by the outside world (e.g., the bleating of the lamb), the creation of language is fundamentally an internal process. If woman has a role to play in the invention of language, it to be the object upon which man focuses his reflection. As Dorothea von Mücke argues, when viewed from a psychoanalytic perspective, the sheep—"white, soft, woolly"—bears a striking resemblance to the mother "as object of tenderness and as object of an aim-inhibited drive."[124]

Herder's *Treatise* ends on a self-reflexive note and with an argument for his own methodology. "How happy he would be," Herder writes, referring to himself,

> if with this treatise he were to displace a hypothesis that, considered from all
> sides, causes the human soul only fog and dishonor, and moreover has done
> so for too long! For just this reason he has transgressed the command of the
> Academy and *supplied no hypothesis*. For what would be the use of having
> one hypothesis outweigh or counterbalance the other? And how do people
> usually regard whatever has the form of a hypothesis but as a philosophical
> novel—*Rousseau's, Condillac's*, and others'? He preferred to work "at *collecting
> firm data from the human soul, human organization, the structure of all ancient
> and savage languages, and the whole household-economy of the human species*,"
> and at *proving* his thesis in the way that the firmest *philosophical truth* can be
> proved. He therefore believes that with his disobedience he has achieved the
> will of the Academy more than it could otherwise have been achieved.[125]

Here, Herder not only recasts the origin of language as human rather than divine, but also repurposes biblical reading practices for his distinctly human project.[126] "The letter kills, but the spirit gives life": these are the words of Paul in Corinthians II, but may as well be lines from Herder, who, despite having "transgressed the command of the Academy" in a literal sense, insists that he has advanced it in spirit. Indeed, not just advanced, but *remade* the Academy's command, rendering it more true and more correct, achieving "more than it could otherwise have been achieved."

In the *Treatise*, Herder becomes the author of everything: the catalyzing question from the Berlin Academy, philosophical truth, the origin itself. The closing paragraph styles Herder's account as unequivocal and all-encompassing—*firm* data, *firm* proof, *all* languages, the *whole* human species. Of all the characters that the text constructs, it is the figure of Herder the author who is most compelling. As man first created language, and through language, a world, so Herder, through an encounter with the "data" of the human soul, creates from within himself the narrative world of the text—the scene of language's origin. The existence of the *Treatise* is a testament to, and a performance of, the same capabilities that supposedly allowed humans to develop language in the first place.

In the passage cited above, in addition to data on "ancient and savage languages" and on "the human soul," Herder lists the "household economy of the human species" as the source of his authority. The foundation of this household economy is of course the nuclear family, the heterosexual couple and child. Sexual differentiation, and the gendered division of labor that it apparently necessitates, is what anchors Herder's tale in the realm of truth, rendering it "firm" and "correct" and allowing it to avoid becoming a mere "philosophical novel." Of all the received ideas that Herder's text interrogates—the link between humans and animals, biblical accounts of language origin, the cultural superiority of the West—the nature of the two sexes is never allowed to come into question, for this "natural truth" serves as his theory's foundation and a justification of his authority.

Fichte, Language, and "Beings of Man's Own Kind"

In a manner similar to Herder's grand self-assertion at the end of the *Treatise*, the German Idealist philosopher Johann Gottlieb Fichte argues that any philosophical investigation into the origin of language must also showcase the author's powers of creation. In "Von der Sprachfähigkeit und dem Ursprung der Sprache" ("On the Linguistic Capacity and the Origin of Language"), the text with which I began this chapter, Fichte writes that the philosopher must simultaneously be the student and originator of language: he "must regard language as being first invented by his own investigation."[127] His account of the creation of language, in other words, should be a performance of this creation itself. In what follows, I read Fichte's theory of the sexes alongside his theory of language origin in order to consider the gendered implications of the narrative world that "Von der Sprachfähigkeit" creates.

ARE WOMEN HUMAN?

On the face of it, we find a clear answer to the question of whether women are human in Fichte's philosophy, an answer that would seem to reject the very question as absurd. In a lengthy section on marriage in his 1797 *Grundlagen des Naturrechts* (*Foundations of Natural Right*), Fichte writes that women, like men, are rational, moral beings. In his *Sittenlehre* (*System of Ethics*) from the following year, women's moral duties are again discussed as part of a larger section titled "Duties of human beings according to their particular natural estate," which includes the relationship of spouses to each other and parents to children. It would, it appears, take a willful misreading to argue that Fichte excludes woman from the category "human being."

And yet the general categorization of woman as rational is belied by Fichte's painstaking efforts to philosophically ground woman's essential difference from man. Woman may be classified as human, but she is a human of a fundamentally different order—we might say that she is less emphatically human than man. As we shall see, what is essential to the human in Fichte's philosophy is precisely what he defines as foreign to woman: complete "free self-activity," which is made possible by reflective, conceptual thinking. Indeed, as Fichte would write early in the *Foundations of Natural Right*, "only free, reciprocal interaction by means of concepts and accordance with concepts, only the giving and receiving of knowledge, is the distinctive character of humanity, by virtue of which alone each person undeniably confirms himself as a human being."[128] Not only does Fichte align woman with passivity (albeit a passivity ostensibly actively chosen), but he also, as Bärbel Frischmann writes, characterizes woman as "'rational' by a natural drive" whereas "man is 'rational' by reason."[129]

Adrian Daub observes that in Fichte's philosophy of marriage, the relation takes primacy over the related terms. "Only by relying on the relationship of *constituens* and *constitutum* to ground the concept of marriage can the model accommodate a striving for *unity*."[130] But the two sexes do not have equal roles to play in the creation of this unity. All humans have a natural drive to propagate the species, Fichte writes. Female sexuality, however, is "the most repugnant and disgusting thing that exists in nature." He continues,

> The lack of a chaste heart in a woman, which consists precisely in the sexual drive expressing itself in her directly, even if for other reasons it never erupts in actions, is the foundation of all vice. In contrast, female purity and chastity, which consists precisely in her sexual drive never manifesting itself as such but only in the shape of love, is the source of everything noble and great in the female soul. For a woman chastity is the principle of all morality.[131]

What Fichte means by *love* I will get to shortly. First, though, a problem: if any expression of the sexual drive is immoral in women, it seems impossible for the natural "propagation of the species" to take place morally. But surely what is natural and what is moral cannot be so misaligned? Fichte solves this problem by arguing that woman's drive is actually not to express her *own* sexuality, but rather to render herself passive *for the sake of man's sexual drive*. Fichte presents this as a fact of nature:

> If the natural drive [to propagate the species] required nothing more than the activity of two people, then our investigation would be finished and there would be no conjugal relationship and no duties pertaining to the same. [...] But things are different in the case now before us. The particular arrangement of nature is such that within the community of the sexes for the purpose of propagating the species only one sex behaves in an active manner while the behavior of the other is entirely passive.[132]

Woman's natural sexual drive is actually not to have a sexual drive at all. This "arrangement" between the two sexes, Fichte writes, gives rise to "the most tender relationships among human beings."[133] Tender though they may be, the characterization of woman as "entirely passive"—while it may solve Fichte's disgust for female sexuality—nonetheless poses a different problem for his philosophical system. For, according to Fichte's own definition, "The character of reason is absolute self-activity";[134] "sheer passivity stands in outright contradiction to reason and abolishes the latter."[135] We are again faced with a conundrum: How can woman, who is apparently a rational being, be passive, if passivity goes against reason itself? Fichte has a response to this as well, although it requires the construction of an elaborate theoretical edifice.

Because humans must possess reason, and reason is active, passivity cannot be woman's *aim*. In other words, a woman cannot want to be passive for herself (she cannot choose passivity because she likes it!), but must desire to be passive to fulfill her husband.[136] Woman's rendering herself passive is what Fichte calls her "love," and her love is what makes marriage possible. A truly "moral" marriage can only take place when the wife's subjection to her husband's will is *die unbegrenzteste* (most limitless).[137] Indeed, Fichte writes repeatedly of woman's "subjugation" and "surrender" in superlative terms:

> Let us consider this first of all from the side of the woman. In giving herself, she gives herself entirely, along with all her powers, her strengths, and her will—in short, her empirical I—and she gives herself *forever*. First of all, *entirely*: she gives her personality; if she were to exempt anything from this subjugation, then what she had exempted would have to have a higher worth for her than her own person, which would amount to the utmost disdain for and debasement of the latter, which is something that simply could not coexist with the moral way of thinking. In addition, she gives herself *forever*, or at least that is what she presupposes. Her surrender occurs out of love, and this can coexist with morality only on the presupposition that she has lost herself completely—both her life and her will, without holding back anything whatsoever—to her loved one.[138]

In return for her unconditional surrender, a husband gives his wife "the most sincere tenderness and magnanimity," and through this reciprocity, the sexual act, "which in itself carries the stamp of animal crudeness," becomes a moral relation.[139]

Fichte's conception of marriage requires woman to surrender herself, her will, and her personality to her husband—what Helmut Müller-Sievers calls her "transcendental disenfranchisement."[140] Not only does woman cede her property and all her rights when she enters into marriage, but she also ceases "to live the life of an individual."[141] Her husband's identity becomes her own, a fact which is "fittingly indicated by the fact that she takes her husband's name."[142] Only by surrendering herself to her husband can woman "[regain] her personality and all of her dignity."[143] The personality she "regains," however, is not her own, but that of her husband. As was the case with her sexual drive, woman's personality and will are defined by their absence.

If woman gives herself up so completely, is she still a rational being? Indeed, can woman even be considered rational prior to marriage, if marriage—and the subservience it entails—is both her natural drive and her "predisposition"?[144] "Man can acknowledge his sexual drive and seek to satisfy it without giving up his dignity," Fichte writes.

> Woman cannot acknowledge this drive. Man can court; woman cannot. [. . .] Reasoning based on the concept of right is of no use here; and if some women are of the opinion that they must have the same right to seek a spouse as men, one can ask them: who is contesting that right, and why don't they therefore avail themselves of it? It is as if one were to ask whether the human being might not have the same right to fly as the bird. Let us, rather, allow the question of right to rest until someone actually flies.[145]

A woman's right to be an individual, to be active instead of passive, is like a human's right to fly: that is to say, a right in name only. While woman could technically make a different choice, this would be a disavowal of her most authentic self. To be moral, woman must marry, and marriage requires passivity. "It is the absolute vocation of every individual of both sexes to enter into marriage," Fichte writes.[146] "A clearly conceived intention never to marry is absolutely contrary to duty."[147] Interestingly, while Fichte at times implies that both men and women are incomplete in their humanity unless they marry,[148] he also accords this complete humanity to man *prior* to marriage. Unlike woman, even before marriage, man can "acknowledge everything that is part of the human being [. . .] [and find] within himself the entire fullness of human nature."[149] Woman, by contrast, exists always as a complement to another. This other being oscillates between the husband and the child, as Stefani Engelstein shows.[150]

In "Some Lectures Concerning the Scholar's Vocation," speeches that Fichte delivered one year before publishing his theory of the origin of language, he explains that the ultimate characteristic feature of rational beings is "absolute unity, constant self-identity, complete agreement with oneself."[151] Man's final end, he writes several pages later, is "to subordinate to himself all that is irrational, to master it freely and according to his own laws."[152] This is, of course, never completely achievable (man is not God, after all), but the endless approximation of this mastery of "all that is irrational" is man's constant striving. Importantly, in Fichte's philosophy, *self-conscious reflection* is the condition of possibility of such a striving, as it is the condition of all moral action. As Michelle Kosch writes about Fichte's theory of moral agency, "the formation of concepts of ends of all kinds and the production of plans for achieving them ('reflection') involves, for Fichte, self-conscious awareness of, and rational evaluation of the consistency of, one's set of motivations, intentions, and beliefs about matters of fact."[153] If free, self-conscious reflection—not to mention *subordination* of irrationality—is a hallmark of the human, how human is woman, really?

Here we get to an important consequence of Fichte's conception of the sexes. Whereas man's relationship to woman is, as Bärbel Frischmann puts it, "the result of a rational decision and a concept," this is not the case for woman. "*Love* is thus the form under which the sexual drive manifests itself in woman," Fichte writes. "But love is self-sacrifice for the sake of another, *not on the basis of a concept, but as the result of a natural drive*."[154] In contrast, "in the man, it is not love, but the sexual drive, that exists *originally*. In him, love is not an original drive at all, but only one that is *imparted* and *derived*, one that is *developed* solely in connection with a loving woman."[155] As a husband witnesses his wife's love for him, he reflects on her sacrifice and chooses to meet her love with magnanimity. Women do not engage in this kind of secondary reflection; they simply are. So Frischmann: "Women do not need their own free will to decide because their nature, as it were, 'decides.'"[156]

This is significant because woman's passive role is not restricted to marriage in Fichte's philosophy. Or, better put, woman's passive satisfaction of her husband through marriage, which is so to speak her telos, means that she can have no individual identity. Because she gives herself up to her husband entirely, she can take part in civic and political life only through him. And because her "decision" to marry is the result of a natural drive rather than a concept, she requires no education—she does not need to develop her capacity for reflective thinking. Moreover, if woman writes, she can only regurgitate what has already been discovered by men, or write about her own feelings. "Thus by virtue of her womanhood," writes Fichte, "woman is already supremely practical, but by no means speculative."[157] This is why women can be "geniuses in the matter of memory," but never in anything that requires conceptual thinking, never "philosophers or mathematical innovators"[158]— and never, we might add, inventors of language.

FROM RECIPROCAL RECOGNITION TO
CHARISMATIC LEADERSHIP

In the previous section, we saw Fichte characterize woman as having a natural drive to render herself passive for the sake of man. While Fichte generally defines woman as a rational being, she possesses a different kind of reason than man: reason by nature rather than reason by concepts. As we turn to Fichte's theory of language, we shall see that this characterization of woman excludes her from participating in the invention of language, an exclusion that the text makes explicit through its examples and its characters. In Fichte's philosophy, woman may be a vehicle for the propagation of language (as she is a vehicle for the propagation of the species), but she cannot be language's originator.

In an investigation into the origin of language, one must "not resort to hypothesis, to an arbitrary list of the particular circumstances under which something like a language *could have* arisen."[159] Here we are at the opening of Fichte's essay, "On the Linguistic Capacity and the Origin of Language," published just one year before the first volume of his *Foundations of Natural Right*. Instead of operating via hypothesis, Fichte proposes to "deduce the necessity of [the invention of language] from the nature of human reason."[160] In the following narrative, he sketches out the creation and development of language, the details of which, he suggests, are not incidental but essential to the story at hand.

Before comparing Fichte's philosophy of language to his philosophy of the sexes, let's review his theory of language origin. Fichte asserts that the human linguistic capacity, or *Sprachfähigkeit*, is "the ability to signify thoughts arbitrarily."[161] To signify thoughts arbitrarily presupposes a choice, which means that language is a voluntary action. Fichte therefore rejects any theory, such as Rousseau's, which claims that language develops from an involuntary eruption of emotion, since this does not afford man the agency he deserves. Once Fichte has established language as voluntary action, however, he is faced with another issue: "In order to voluntarily decide to invent a language, an idea of it would have to be presupposed. Hence the question: How did the idea reciprocally to share their thoughts through signs develop in human beings?"[162] Indeed, the text asks, "is there in human nature the means which would necessarily have to be available in order to realize the idea of a language?"[163] In order to answer these questions, Fichte takes his reader through a series of logical steps that requires him to define the human in essentialist terms. What emerges from Fichte's account of primordial man is a vision of the human that is defined by activity and reason.

"It is fundamental to man's very essence that [man] seeks to subjugate the power of nature," Fichte writes.[164] Primordial man did this by digging caves,

clothing himself in natural materials, and making fire. When it was not possible for man to conquer nature, he ran from it; man and nature were thus
in constant warfare.[165] Why does man not act the same way with his fellow
men? That is, why are men not constantly trying to vanquish each other? In
attempting to modify "raw or bestial nature according to his own purposes,"
Fichte writes, man seeks to bring nature into harmony with himself—he seeks
to make nature *reasonable*.[166] Such a modification is not necessary when he
encounters another human being. Man recognizes another man as a "being
attuned to himself,"[167] a "being of his own kind."[168] Nonetheless, it is not immediately smooth sailing in the first human encounter, because there remains
the possibility for misunderstanding and thus for hostility. Hence the need
to communicate. From here, the idea of language "awakens" within man.[169]

Fichte's account of the invention of language is premised upon two humans' recognizing each other as human. Is woman also a "being of man's own
kind"? To a text interested only in the *necessary* development of language,
we might ask: Are the protagonists at the center of this narrative *necessarily*
sexed? On the one hand, in this very first scene, the text does not explicitly mark the characters as male. Furthermore, the first human encounter is
grounded in reciprocity, which resonates with Fichte's description of marriage. We may remember that Fichte described marriage as a relationship in
which woman gives herself to man and man gives her "magnanimity and
tenderness" in return. Given that the (heterosexual) marital pair is such a
fundamental relational model in Fichte's philosophy, one could imagine that
it might serve as the basis for conceptualizing other dyadic relations.

On the other hand, there is a major difference between the scene Fichte
portrays here and his portrayal of marriage. When two primal humans encounter each other in Fichte's origin of language essay, they enter into a relation of sameness: like recognizes like. Here is Fichte reflecting on how one
human recognizes another as a "being attuned to himself":

> Only a being which, after I have expressed my purpose to it, alters its own
> purpose in relation to my expression, which, for example, uses force if I use
> force against it, which acts kindly to me if I act kindly to it: only such a being
> can I know to be rational. Then I can conclude from the reciprocity which has
> arisen between it and me that it has comprehended a representation of my
> manner of behavior, has adapted it to its own purposes, and now as a result of
> this comparison freely gives its own actions another direction.[170]

Fichte's definition of marriage, however, is based on a relation of fundamental
difference. If woman is to act in accordance with her natural drive, she cannot
meet man's force with force, as she cannot meet man's activity with activity.

That language is fundamentally an *act* for Fichte—and a voluntary act, to boot—already aligns language with man rather than woman, as woman's most moral act is the act of rendering herself passive. Moreover, if we look closely at the passage quoted above, we see that the initial encounter Fichte describes requires both humans to engage in conceptual thinking—the two characters "conclude" and "comprehend." Woman, as we may remember, is disallowed this kind of conceptual reflection. While the masculine origin of language is only implicit in this first scene, it becomes explicit in the narrative that follows.

According to Fichte, the *Ursprache*, or the first language invented by humans, was a "hieroglyphic language."[171] The signs comprising the Ursprache were not arbitrary, but imitated the sounds of nature: "the lion, for example, was expressed through imitating its roar, the wind through imitating its howling."[172] This kind of language worked fine when there were just two people and they were in close proximity; speakers facilitated understanding by pointing and gesturing. But if two men were far away from each other (say, on a hunt), or if there were a number of people assembled, this kind of language proved insufficient.[173] Thus hieroglyphic language came to be replaced by what Fichte calls "audible language,"[174] which uses arbitrary signs and which is the kind of language that we know today. As Fichte describes it, the transition from hieroglyphic to audible language is a multistep process. Most interesting for our purposes is the way that Fichte's narrative depends on the innovations of two male characters: the tribal leader and the father.

Fichte imagines that a charismatic leader, skilled in the art of speaking, could have introduced the practice of using arbitrary signs:

> For a people which—as is known from wild clans—loves assemblies, works and feasts socially, and so forth, it is easy to imagine that one man, by virtue of his mental [or spiritual, *seines Geistes*] superiority, makes a proposal before the others and, without being elected for this, serves as the military leader in war and the speaker in their gatherings. Such a man, whose speeches one most carefully attends, will by habit attain a skill in speaking and, by virtue of this skill [or fluency], it so happens that although he signifies things only casually, it is not taken as inappropriate if he skips over this or that sound in speaking. One will soon become accustomed to this deviation and will easily learn to understand this casual sort of signification. He will gradually get further and further away from the exact imitation of the natural sounds, his signification will become little by little more casual [or fluent], briefer, and easier.[175]

This more "casual" way of speaking spreads because others find it convenient. Fichte then tackles another question: Once the idea of using arbitrary signs had

been instituted, how were more words created and spread to others? The answer to this lies in the domestic household. "By its very nature," Fichte writes,

> this would have to be the business of the mother and father of a family. In their
> domestic affairs they would often have occasion to invent many new sounds
> whereby they could instruct the members of their household in how to deal
> with an object by means of an expression which they first explained by exhib-
> iting the object itself. Through constant usage these expressions became easier
> for the mother and the father themselves.[176]

Although the above passage seems to suggest that the development of language is a mutual project between husband and wife, this is the last we hear of
the mother. She is quickly excised from the narrative, as Fichte's subsequent
examples make clear that it is the father, specifically, who is doing the important work of fabricating words.

Like Herder, Fichte characterizes language as a kind of patrimony. Not
only do patriarchs pass on language to their male progeny, but they also be-
queath it to the rest of their tribe. Men who are smart enough to concern
themselves with the invention of language, Fichte writes, will also gain influence over their social group, and introduce the words they invent at deliberative meetings: "In this way the invention of a father will soon spread through
the entire tribe."[177] That such a tribal leader is assumed to be male hardly
needs mention. Not only is the figure described as a father and a potential
military leader in war,[178] but Fichte's conception of sexual difference pre-
cludes women from taking part in civic life. Whether in "primordial" tribes
or modern Germany, women cannot attend or speak at deliberative meetings.
According to Fichte, women have no independent political voice; they take
part in the collective only through the participation of their husbands.

I have analyzed Fichte's characters in such painstaking detail not to prove
the chauvinism of a centuries-dead author but to show how and why his theory excludes women from the origin of language—as well as the consequences
of this exclusion. Better known than Fichte's "Linguistic Capacity" essay are
his *Addresses to the German Nation*, which he gave in Berlin between 1807 and
1808 in order to, in the words of Andrew Fiala, "inspire the German people
to struggle against the tyranny of French occupation."[179] As Fiala emphasizes,
while Fichte's focus on the political dimension of language in the *Addresses* is
largely absent in the "Origin" essay, he maintains the same general conception
of language in both texts, namely that "the transcendental ground of language
is the intersubjectivity of human freedom."[180] In the fourth *Address*, Fichte
famously describes German as unique among languages. Germans, he claims,
retained and developed "the original language of the ancestral race," whereas

other "Teutonic" tribes "adopted a foreign language and gradually modified it after their own fashion."[181] This is important for Fichte because he argues that language and *Volk* are intimately connected. He defines a nation not by the land its people inhabit—this is "quite insignificant"—but by a people's relationship to its language.[182] In this way, Fichte asserts an extreme form of monolingualism, where only the people of a language's "ancestral race" can experience an immediate resonance between their language and their identity. German is one such language, a language that

> from the moment its first sound broke forth in the same people, has developed uninterruptedly out of the actual common life of that people; a language that admitted no element that did not express an intuition actually experienced by this people, an intuition that coheres with all the others in an interlocking system. Let the original people who spoke this language incorporate however many other individuals of another tribe and another language: if these newcomers are not allowed to raise the sphere of their intuitions to the standpoint from which henceforth the language will continue to develop, then they remain without voice in the community and without influence on the language until they themselves have gained entry into the sphere of intuitions of the original race. And so they do not form the language but the language forms them.[183]

Speaking an "original" language is not merely a matter of retaining words or grammatical forms from prior centuries. Instead, it has to do with a people's stance or activity within their language, what Étienne Balibar calls their "moral action in language." For Fichte, writes Balibar, "the originary does not demonstrate *whence a people comes*, but *what it is moving toward*, or still more precisely the moral destiny that it actively gives itself, whose proof Fichte thinks he can locate in a particular 'German' disposition to *take seriously* the words of the language: to 'live as one speaks' and to 'speak as one acts.'"[184] Speaking an original language, in other words, allows the human to live most authentically, to "mak[e] the idea penetrate into life."[185]

What Fichte says about "newcomers," or people who adopt the language of another tribe, may equally be said about women: "they do not form the language, but the language forms them." Women necessarily cannot participate in the "reciprocal belonging"[186] of language and people because women do not actively create, only passively receive. Moreover, unlike foreign "newcomers," who ostensibly can assimilate and "gain entry into the sphere of intuitions of the original race,"[187] women always maintain their essential alterity. In this, their situation is like that of another marginalized figure, "the Jew," at least as Richard Wagner would infamously describe it several decades later in "Das Judenthum in der Musik" ("Judaism in Music"). "The Jew speaks the

language of the nation in whose midst he dwells from generation to generation, but he speaks it always as an alien," Wagner insists. "The general circum-
stance that the Jew talks the modern European languages merely as learnt,
and not as mother tongues, must necessarily debar him from all capability of
therein expressing himself idiomatically, independently, and conformably to
his nature."[188]

Yet while "the Jew" in Wagner's anti-Semitic diatribe has his own—
allegedly inferior—language of Yiddish, "the woman" in Fichte's imagination
has no national language of her own. As Adrian Daub observes in his reading of Fichte's 1794 lecture "Concerning Human Dignity," women occupy a
unique position: their exclusion from all emphatically human activity is not
a result of historical circumstances but "a natural moral fact."[189] In comparison to so-called "primitive" peoples, whose "submissions are social," woman's
submission to man is for Fichte "absolute and necessary." So does "the trium-
phal march of human activity in the world ha[ve] as its obverse a naturally
mandated lack of dignity that consigns half of humanity to an eternally sub-
altern position."[190]

When Fichte correlates language with national/cultural identity in the
Reden, he performs a gesture that is typical of nineteenth-century linguistic
thought: language and Volk are mutually affirming. When we press Fichte's
philosophy, however, we see that "woman" poses a dilemma to this model of
linguistic nationalism. Women constitute a distinct culture within a nation's
monoculture; because of their "natural" differences, they can never be fully
assimilated. Yet since women play a necessary reproductive role in *sustaining*
the monoculture, they also cannot be exiled or expelled, can't be conquered
militarily like Napoleon's army. The institution of marriage curbs woman's
difference by effacing it, aligning her with the identity and values of her husband. Yet in marriage woman's alterity only lies dormant. What if women
took up the rights accorded to them in name only? As sure as Fichte's texts are
about the model of sexual complementarity he advances—it is natural, moral,
and right—we can also sense an anxiety evidenced in the elaborate theoretical system he must devise to contain and explain away woman's function in
sexual and linguistic production.

Wilhelm von Humboldt and the Matter of Form

While Wilhelm von Humboldt is today best known as a linguist and educa-
tional reformer, he started his career as a philosopher of sexual difference.
In the 1790s, Humboldt published two essays in Friedrich Schiller's journal
Die Horen: "Ueber den Geschlechtsunterschied und dessen Einfluß auf die

organische Natur" (On the division of the sexes and its influence on organic
nature) and "Ueber die männliche und weibliche Form" (On male and female form). In these texts, Humboldt characterizes femininity and masculinity as complementary principles whose unification leads to ideal beauty and
ideal humanity. Humboldt defines femininity as passive matter, masculinity
as active form. "Here is where the difference between the sexes begins. The
generative power is more attuned to activity, the receptive power more to
reaction," he writes in "Ueber den Geschlechtsunterschied." "What is enlivened by the former we call male, what is animated by the latter we call female.
Everything male exhibits more self-activity, everything female more passive
receptivity."[191] Humboldt reinforces this distinction throughout the two texts.
Masculinity corresponds to the active "power of life," while femininity corresponds to a receptive "abundantly overflowing fullness, too bountiful to
be able to be stimulated by its own strength."[192] Male beauty is marked by a
"supremacy of form," while female beauty is "more feeling through the free
fullness of material."[193]

Humboldt scholars have disagreed about whether his gender binary is
also a hierarchy, and whether he understands gender to be divorced from sex.
Simon Richter and Peter Hanns Reill both suggest that Humboldt establishes
masculinity and femininity as equal principles, and that *gender* does not necessarily mean *sex* in Humboldt's texts.[194] It is true that Humboldt insists that
the unification of the genders leads to a genderless perfection, what he calls
the "ideal of pure and sexless humanity."[195] It is also true that, while Humboldt asserts an absolute difference between femininity and masculinity,[196]
he maintains that this is a difference of *direction* rather than of ability.[197] Nevertheless, although the abstract unification of the genders is what interests
Humboldt, his characterization of masculinity and femininity depends on
asserting a fundamental difference between the sexes as biological fact. As
Christina von Braun, Catriona MacLeod, and others have shown, this leads
Humboldt to establish an implicit hierarchy between the genders and to tie
gender ineluctably to sex.[198] A closer look at three aspects of Humboldt's construction of gender will elucidate this further.

First, there is the biology of procreation. Humboldt takes his characterization of the masculine and feminine principles from a classical understanding
of the male and female roles in sexual reproduction. According to this model,
which can be traced back to Aristotle, the man contributes the "principle of
movement" necessary for reproduction, while the woman provides the material in which this reproduction takes place. Or as Aristotle puts it in *On the
Generation of Animals*, the male is "that which generates," while the female
is "that out of which it generates";[199] "the female, qua female, is passive, and

the male, qua male, is active."[200] Although Humboldt considers *Männlichkeit* and *Weiblichkeit* in an abstract sense, their relationship to sexed bodies is not arbitrary. On the contrary, it is precisely the sexed body from which his conception of gender derives.

Second, even if Humboldt describes ideal humanity and ideal beauty as gender*less*, the ideal can only ever be located within a male subject, as Catriona MacLeod has argued.[201] This is evidenced not only implicitly, by his exclusively male examples (he counts Homer, Dante, and Aristotle among the geniuses who merge "masculine reason" with "female fantasy"), but also explicitly, by his description of men's and women's divergent capacities.[202] Man, Humboldt writes, can "renounce his sex, as it were, and go beyond his naturally-determined purpose; that is, accomplish more than even what his highest purpose commands." Woman, on the other hand,

> must preserve every female characteristic [*weibliche Eigenthümlichkeit*] with gentle care so as to not destroy that living expression of its very shape; and if this effort completely fails, she will merely sink to her natural purpose and the performances of external everyday life, or transition to occupations that do not actually belong to her sphere. For here, too, as soon as one leaves the boundaries of mere natural ends, femininity is created to give only the highest, and whoever turns to her with demands only proves his ignorance of the gender.[203]

Only man can move beyond his sex to approach the ideal, whereas woman must, in MacLeod's words, "remain within the bounds dictated by biological difference and [serve] willingly in the cause of male completion."[204] Woman's relation to *Bildung* is to help man achieve it, to act as a counterbalance to his activity and rationality.[205]

Additionally, Humboldt's characterization of femininity and masculinity is essentially identical to his characterization of women and men. In "Ueber den Geschlechtsunterschied," Humboldt uses the terms *man* and *masculine* interchangeably, referring, for instance, to "man and his sex."[206] What's more, as Braun also points out, Humboldt's "Plan for a Comparative Anthropology"—a text contemporaneous with the *Horen* essays—uses the same language to describe women as he did femininity. "As with men the mind," Humboldt writes,

> so with women the sentiments are the most animated and active [. . .] the important difference is that women are the receiving and preserving part, that only they have their completely own feeling of being a mother, and that the character of their sex is generally more intimately woven into their personality.[207]

Just as he claimed about the "feminine principle," Humboldt characterizes women as receivers and preservers. And whereas femininity was correlated with materiality in the *Horen* essays, so too are women defined here by their corporeality, their capacity for motherhood.

Finally, Humboldt's texts display a partiality toward "the masculine" in their very genre. One does not philosophize in "receptivity" and "materiality." It is form that brings forth order, meaning, and reason, and that makes it possible to craft a theory of gendered aesthetics in the first place. As I will argue in the following section, Humboldt's theory of language betrays a clear valorization of activity and form—qualities that belong, by his own categorization, to the masculine side of the gender binary.

SPRACHFORM AND MASCULINITY

Although Humboldt does not narrate the development of language like Herder or Fichte, his writings have much to say about what language is, how it originated, and how it should be studied. One of the most comprehensive records of Humboldt's philosophy of language is found in *Über die Verschiedenheiten des menschlichen Sprachbaues und ihren Einfluß auf die geistige Entwickelung des Menschengeschlechts* (*On the Diversity of Human Language Construction and its Influence on the Mental Development of the Human Species*), which he wrote as the introduction to his study of the Kawi language of Java, and which was first published posthumously in 1836.[208] While the origin of language is not his primary focus here, Humboldt makes claims about the general conditions that make human language possible. He rejects, for instance, the idea that language developed out of a "need for mutual assistance" (as, for instance, Condillac or Smith had argued).[209] Humboldt likewise repudiates the idea that language began with the designation of a few objects by specific words, which were then strung together to create speech. Instead, Humboldt argues that words emerged "from the totality of speech"[210]— language was always already a complete organism.

Furthermore, in contrast to a theory that would place man in an original state of nature, slowly shedding his animality as he invents language to compensate for his helplessness, Humboldt emphasizes an original relationship between language and thought. Following Herder, he writes that speech is a necessary condition for thinking, necessary even for the individual in solitary seclusion. "In appearance, however, language develops only socially," Humboldt continues, "and man understands himself only once he has tested the intelligibility of his words by trial upon others."[211] According to Humboldt, this development of language follows two main stages. First there is the period in which "the sound-making impulse is still in a state of growth and lively activity." Then, once a language's form has been shaped, "a seeming halt occurs, and there then follows a visible decline in that creative sensuous impulse. Though even from the period of decline it is possible for new life-principles and novel transformations of the language to emerge."[212] That is to say, the invention of language is a never-ending, continuous process.

Though Humboldt devotes the entirety of the *Kawi* introduction to outlining his theory of language, even by the end of the text, his conception of language remains rather opaque. Humboldt defines language as an "organism" and a "formative power," a "necessary condition for thinking" and the "outer appearance of the spirit of a people," and—most famously—"no product [*Ergon*], but an activity [*Energeia*]." Humboldt's philosophy is notoriously dense and requires slow parsing. I will discuss these claims, as well as their standard interpretations, in more detail shortly.

First, however, I would like to consider a few emblematic examples of how Humboldt's interest in gender has been understood to relate to his work on language. The prevalence of gendered categories in Humboldt's thought beyond the *Horen* essays has been well noted, though it is usually read as a stand-in for other issues. Peter Hanns Reill, for instance, writes that although Humboldt employed a "variety of dyads" in his work on language and history, he favored the dyad masculine/feminine and subsumed all other categories within it.[213] Reill explains Humboldt's preference for the masculine/feminine binary via recourse to biography, suggesting that it "reflect[s] some of Humboldt's own personal sexual problems," in particular his ambivalence "in his male/female relations, sometimes dreaming of being controlled, other times indulging in sadistic fantasies of control and enslavement."[214] On the other hand, Jürgen Trabant, who has written extensively on Humboldt's philosophy of language, argues that Humboldt's deployment of a "model of synthesis developed out of the sexual union" should be seen as a response to Kant and his theory of the conditions of possibility of knowledge";[215] in Humboldt, the subject-object problem and the problem of imagination are thought of in terms of a love relationship.[216] Trabant also suggests that Humboldt's use of the sexual union as a model for discussing politics may be rooted in his reaction to the French Revolution.[217] James Underhill, on the other hand, contends that the sexual metaphors that "structure Humboldt's thought and the vocabulary that colours it are often reshaped from the words, concepts and metaphors common to the discourse of the Zeitgeist that Goethe and Schiller incarnated."[218] I aim to challenge this scholarly dismissal of the significance of Humboldt's theory of gender for his theory of language, showing that Humboldt characterizes language in its quintessence such that it corresponds rather remarkably to his definition of masculinity. The binary distinctions that Humboldt employs to describe language in its most emphatic and authentic sense—*form* rather than material, *activity* rather than product—rely on the gendered framework developed in the *Horen* essays, which, similar to Fichte, establish masculinity as self-activity and femininity as its receptive opposite.

If, as Humboldt contends, the bringing forth of language is the true nature and "inner need" of human beings; if language is necessary not only for the development of mental power but also for a "world-view"; if a language's distinct form is what makes a Volk a Volk: Then what is the role of woman in all this? In practice, Humboldt excluded women from the project of Bildung, and thereby from any role in institutionalized civic life. When he reformed the school system as head of the Prussian educational administration, he "made no reference to any formal education for girls beyond the elementary level" and "did nothing to create girls' secondary schools during his brief but crucial tenure" in the early 1800s.[219] Insofar as it relies on gendered categories to define what language is and is not, Humboldt's philosophy of language not only reinforces but actually enables these exclusions. If woman is misaligned with language, she is misaligned with everything in which language is implicated: thinking, creativity, genius, communal identity. This misalignment means that woman cannot herself form or create language but can only be the *vehicle* of its propagation. Or as Friedrich Kittler would argue in his study of language pedagogy around 1800, in the production of language, woman has a most singular function: gestation.[220]

OUR LANGUAGES, OURSELVES

Two characterizations of language stand out most prominently in the *Kawi-Einleitung*: language as activity and language as form. Humboldt repeatedly stresses that language must be understood as an activity: "To describe language as a work of the spirit is a perfectly correct and adequate terminology, if only because the existence of spirit as such can be thought of only in and as activity." "Language proper" (*die eigentliche Sprache*), Humboldt furthermore explains,

> lies in the act of its real production. It alone must in general always be thought of as the true and primary, in all investigations which are to penetrate into the living essentiality of language. The break-up into words and rules is only a dead makeshift of scientific analysis.[221]

We use the term *language* to talk about grammar, vocabulary, written words, but according to Humboldt, this is not language in its truest sense, not "language proper." True language is found in speech, in the moment of its creation. Humboldt distinguishes between these two notions of language, which we could call the authentic and the deficient, throughout his text. "We must look upon *language*, not as a dead *product*, but far more as a *producing*," he argues.[222] Even writing is just "an incomplete, mummy-like preservation, only needed again in attempting thereby to picture the living utterance. In itself it is no product [*Ergon*], but an activity [*Energeia*]."[223]

What exactly Humboldt means by this formulation has been the topic of much debate. Hans Aarsleff, who describes the statement as being "the subject of endless and aimless speculation without ever leading to a viable answer," argues that Humboldt borrows the concept of Energeia from French Enlightenment thinkers such as Diderot and Beauzée.[224] Other critics claim that it follows from his engagement with Aristotle (although they disagree about the source text), yet others from his reading of James Harris.[225] I am less interested in debating Humboldt's influences than in investigating the consonance between language-as-Energeia and masculinity, which has been widely neglected in the existing scholarship devoted to his work. To understand language as Energeia means to understand it as the "formative power" that makes cognition possible. Language is the "animating breath" that transforms the world into thought.[226] While this description holds true for all languages, each language performs the transformation differently, according to its *innere Sprachform*.

When Humboldt uses the term *form*, he does not mean grammatical form. In fact, the term does not refer to a particular aspect of language that can be easily pinpointed or identified. A language's Sprachform extends, Humboldt writes, "far beyond the rules of *word-order* and even beyond those of *word-formation*."[227] Instead, it is something larger, which permeates a language entirely, down to its smallest elements, and in this way distinguishes it from all other languages. Because language and thought are inextricable, the "mental individuality" of a people and their "language shape" (*Sprachgestaltung*) stand in "intimate fusion with one another."[228] We thus have linguistic diversity among mankind because and insofar as the mental individuality of nations is itself different.[229] Here, we arrive at Humboldt's famous conception of language-as-worldview. The spirit and language of a people mutually reinforce each other; "we can never think of them sufficiently as identical."[230] This, again, is an identity based on *form*—"Intellectuality and language allow and further only forms that are mutually congenial to one another."[231]

The *Horen* essays established a model of sexual reproduction in which masculine form activates receptive feminine matter. This binary also permeates Humboldt's theory of language. A language's form "is contrasted, indeed, to a *matter*," Humboldt writes,

but to find the matter of linguistic form, we must go beyond the bounds of·
language. Within the latter it is only relatively speaking that one thing can be
regarded as the matter of another, e.g., the basic words in contrast to declension. But the matter here is again perceived in other connections as form. [. . .]
In an absolute sense there can be no *formless matter* within language, since
everything in it is directed to a specific goal, the expression of thought, and
this work already begins with its first element, the articulated sound, which
of course becomes articulate precisely through being formed. The real matter
of language is, on the one hand, the sound as such, and on the other the totality of sense-impressions and spontaneous mental activities which precede the
creation of the concept with the aid of language.[232]

Humboldt presents the form/matter binary in terms of a hierarchy, stressing
that while form is inherent to language, matter is something external. Or to
put it differently, form is what makes language *language*, what distinguishes
language from its constitutive parts (sounds, sense impressions), which comprise it but do not define it. Although language ostensibly requires both matter and form to exist, since form must form *something*, it is noteworthy that
Humboldt only makes the caveat that there "can be no *formless matter* within
language," and not vice versa.

Because form is privileged by language itself, it must also be privileged
by those who study it. The study of languages can only truly be an investigation into the "identity and affinity of their *forms*."[233] This investigation has
its limits. The study of language, as the study of humanity, ultimately leads
to the impossible task of justifying human nature. Indeed, Humboldt makes
this impasse clear, writing that language in its most fundamental sense, "the
inseparable bonding of thought, vocal apparatus and hearing is unalterably
rooted in the original constitution of human nature, which cannot be further explained."[234] However scholars dismember and dissect a language, then,
"there always remains something unknown left over in it"—and this is precisely "wherein the unity and breath of a living thing resides."[235]

Even though we may never know a language's Sprachform completely,
Humboldt is clear that language study is nonetheless worthwhile. It allows us
to obtain the understanding necessary to gain a general "survey."[236] Moreover,
studying language form is important because it gives us insight not only into
specific languages and their speakers, but also into what it means to be human more generally. Discussing the relationship between language, the human, and nature, Humboldt writes that language "does not merely implant an
indefinable multitude of *material elements* out of nature onto the soul,"

it also supplies the latter with that which confronts us from the totality as *form. Nature* unfolds before us a many-hued and, by all sensory impressions, a diverse manifold, suffused with a luminous clarity. Our subsequent reflection discovers therein a *regularity* congenial to our mental form. Aside from the bodily existence of things, their outlines are clothed, like a magic intended for man alone, with external beauty, in which regularity and sensory material enter an alliance that still remains inexplicable to us, in that we are seized and carried away by it. All this we find again in analogous harmonies with language, and language is able to depict it.[237]

Mental activity, nature, and language mutually reinforce each other; they are in "analogous harmony." Thus, in language, humans find a "congenial" affirmation of themselves and the world they inhabit. Notice the way that Humboldt de-emphasizes materiality here: "Language does *not merely* implant [. . .] material elements"; "*aside from* the bodily existence of things." It is their similarity as *form* that allows for the resonance between mind, nature, and language. As Humboldt continues this section, he again stresses the importance of form: "The regularity of language's own *structure* is akin to that of nature, and in thereby arousing man in the activity of his highest and most human powers, it also brings him closer, as such, to an understanding of the *formal* impress of nature."[238] "Man is a prisoner of language," writes Lia Formigari, summarizing Humboldt's position. "He can escape only into another language. But this kind of linguistic solipsism is offset by a sort of linguistic occasionalism. The prisons of *Weltbilder* are breached by the notion of analogy, the universality of human nature."[239] The form of language, in other words, "is not foreign to the true original nature of man."[240]

Woman's status in relation to this true and original nature is ambiguous. It is form that activates our "highest and most human powers," yet form is explicitly masculine in Humboldt's thought. Women would seem to be excluded from experiencing such a "congenial" harmony between self, language, and world, since their selves are aligned—over and against form—with materiality. To be sure, as we saw in the *Horen* essays, although the union of form and matter, masculine and feminine, is the ultimate ideal, this union can take place only in the man. Just as the subject of Bildung is specifically masculine, so too is the speaking subject, the collective *we* referenced in the quotation above. It is true that Humboldt does not make statements that directly exclude women from these essential human experiences, but that is precisely the point. The masculinity of the universal subject does not need to be stated outright because—for Humboldt's text, at least—it is impossible to imagine it otherwise. Although Humboldt does not mention woman's relation to language here, he does include a section on "women's language" in an earlier version of the *Kawi-Einleitung*. I discuss this in more detail in chapter 2, but for now it suffices to say that *Weibersprache* receives attention because it is an aberration, a deviation from "regular" language, which is coterminous with "men's" language. Predictably, there is no analogous section on *Männersprache* (men's language) in Humboldt's work.

While it is hardly a revelation to say that the universal subject is in fact restricted, it is nevertheless worth considering how this operates in Humboldt's text. The history of misogyny is not a history of monotony, at least not in the means by which it is produced. Authors of earlier theories of language often also asserted that women spoke differently—and less rationally or truthfully—than men. Rousseau, for instance, states in the *Émile* that women have developed to have "flexible tongues; they talk sooner, more easily, and more attractively than men. [. . .] Man says what he knows, woman says what pleases. He needs knowledge to speak; she needs taste."[241] In Rousseau, there are two kinds of language; every description of women's speech finds its counterpart in the description of men's. In Humboldt, the difference of sex in language is a difference specific to women alone. There is no need for the category *men's language* because language as such is already masculine.

For a long time, we have known the alignment of language and masculinity to be true in practice, borne out by the manifold restrictions on women's speech. We might think of the French Revolution, which created, as Joan Wallach Scott writes, "an inherent conflict between principle and practice." While the Declaration of the Rights of Man and of the Citizen "succeeded in rallying patriots to the Revolution," it also "made possible the discontent of those (women, slaves, and free men of color among them) who were excluded from citizenship and by the terms of the constitution promulgated two years later."[242] In contrast to texts that posit a genderless universal in the abstract, only to restrict it to the masculine in practice, what Herder suggests, and Fichte and Humboldt systematize, is a *theoretical* masculinization of language. *Theoretical* has a double meaning here. These authors suggest that if women attempt to engage in theoretical reflection (that is, if they attempt to become philosophers), this degrades both women and philosophy. When language is understood to be rooted in the conceptual—whether this is a reflective taking-awareness, recognition, or mental form—to exclude women *from* the theoretical also means to exclude them *in* the theoretical: from theories of language's necessary origin. This theoretical limiting of possibilities both relies upon and reinforces a pernicious logic of exclusion. If woman cannot represent the *homo loquens*, she is excluded not only from linguistic texts and the language disciplines but also from intellectual and political life, and even from the category *human*.

*

The alignment of language with masculinity, which I traced in the sections above, reverberates through the German philosophy of language in the nineteenth century. In the 1808 *Über die Sprache und Weisheit der Indier* (*On the Language and Wisdom of the Indians*), for instance, Friedrich Schlegel distinguishes between two types of languages, those with inflection and those without. According to Schlegel, inflected languages, such as Sanskrit, are of a higher order and arise neither out of instinctive cries nor out of the imitation of nature, but result from "the most profound study and the clearest thoughtfulness."[243] Even in their most original forms, such languages are the result of labor and reflection, exemplifying "the loftiest ideas of the pure world of thought."[244] Incidentally, these are the same attributes that Schlegel elsewhere ascribes to the masculine contra the feminine.[245]

Jacob Grimm similarly asserts an implicitly male origin to language in an essay from 1851. Like Herder, Fichte, and Humboldt, Grimm argues that language is a distinctly human phenomenon, neither given by God nor developed out of animal sounds. Unlike his predecessors, however, Grimm considers the effect of women in the trajectory of language's development. Describing a historical period in which several generations have already come into existence, he writes that "one can attribute to women [. . .] their own characteristics, some of which have been adopted from men into a separate custom and function, even peculiarities of dialect for expressing the terms that they are especially familiar with from an early age."[246] Although Grimm ascribes women a role here, he notably does not portray them as the first human speakers, situated at language's *origin*. Instead, women influence an already created language, impacting the development of different grammatical forms. As Grimm will argue in his study of grammatical gender, feminine linguistic forms, like women themselves, are by nature derivative.

Grimm originally delivered his reflections on the origin of language as a lecture at the Royal Academy of Sciences in Berlin. Reprinted multiple times, this essay reached a large audience. One reader was the *Oberlehrer* of a gymnasium in Brieg (now Brzeg, Poland), whose contribution to the school's 1853 *Einladungs-Programm zur Oster-Prüfung der Schüler aller Klassen* (Invitational program to the Easter exam of students of all classes) was a treatise consisting of "comments and additions to the writings of Grimm and Steinthal 'on the origin of language.'" In the scope of its influence, however,

Grimm's text pales in comparison to Herder's *Treatise*, which was discussed by his epigones for over a century.[247]

Indeed, the *Treatise* was not only quoted and copied, even while Herder himself modified his theory of language over the course of his career,[248] but the claims articulated by his text—the mutual dependency of language and reason, language as defining the human, the symbiotic relationship between language and culture—remained influential, even as conjectural origin narratives declined in popularity. These claims are slow to lose the gendered sediment that accrued in their original context of production.[249] We see this, for example, in the work of Heymann Steinthal, the "Steinthal" mentioned by the Oberlehrer above. Steinthal was a linguist, philosopher, proponent of folk psychology, and inheritor of the Humboldtian tradition. He published a number of texts dealing with the origin of language, including his 1851 *Der Ursprung der Sprache im zusammenhange mit den letzten Fragen alles Wissens* (The origin of language in connection with the ultimate questions of all knowledge) and 1871 *Abriss der Sprachwissenschaft* (Outline of language science). In the *Abriss*, he writes that the emergence of language is like the fertilization of an embryo, while its development is like the embryo's subsequent formation.[250] As Steinthal has it, man is not the *creator* of language, but the vehicle through which language creates itself: "Neither primitive man nor the child of later generations makes or creates language; rather, it arises and grows in man, he gives birth to it."[251] Lest one think this would place man in a passive, feminine position, Steinthal clarifies: "Once he has given birth to language, he has to take up his own birth, to learn to understand."[252] In giving birth to language, the human also gives birth to himself. He is at once matter and form, father, mother, and child: a subject completely coherent and self-sufficient.

Why do theories of language craft metaphors, construct characters, propose narratives? Why, in other words, does a philosophical discourse take recourse to literary convention—especially when its authors repeatedly insist that they are *not* in the business of fiction? In the texts examined in this chapter, literary devices provide an organizing structure and sense of causality. Developmental narratives obscure the unknowability of language's origin, overwriting it. If the origin of language is fundamentally irrecoverable—since the origin of language is coeval with the origin of the human as such—then narrative allows for a point of beginning to be posited, which in turn enables a concept of development. And if language is ultimately unknowable in an absolute sense, as Humboldt for instance suggests,[253] then describing language through metaphor—like that of form versus matter—allows a text to set aside this lacuna and discuss language in definitive terms, nonetheless.

In the absence of total knowledge, literary figures and modes allow for the semblance of a unified and all-encompassing theory.

That their theories rely on literary modes is not something that Fichte, Herder, or Humboldt acknowledges. They claim to describe not what is possible, but only what is entirely true, how things "must have happened." One way that their texts transform fictional conjecture into philosophical certainty is by the accordance of this conjecture with the fundamental "truth" of sexual difference. The natural "fact" that masculine activity is opposed to feminine passivity, that the heterosexual family is structured patriarchally, that woman's function is to be man's helpmate: these gendered assumptions ground a text's narratives and metaphors, making them appear self-evident. They make possible the text's imagined coherence, its claims to necessity.

Incoherence and contingency, by contrast, are the domain of woman. Not only is women's speech *characterized* as incoherent, but this characterization of woman is also a *site* of logical incoherence within the texts. Required for reproduction, but not the agent of reproduction, designated human, yet not as sufficiently human as man, woman's ontological status remains unsettled and context-dependent. In this way, woman's unstable position elegantly undermines the absolute knowledge and totalizing systems that theories of language claim to produce. Insofar as the female subject's contingent relation to the category "human" plays a constitutive role in the creation of "necessary" narratives about language and its origin, such philosophical texts end up incorporating the very contingency that they seek to exclude. Thus does the theoretical masculinization of language create not only a political problem (if language is the arbiter of the human as such, to align language with masculinity means to dehumanize woman) but also a philosophical one as well. The theories of Herder, Fichte, and Humboldt cannot dispense with woman altogether, but they also do not allow her to be a full subject. Despite attempts to fix woman's nature philosophically, she remains suspended between animal and human, "primitive" and "civilized," occupying an always-shifting position depending on the topic at hand.

Herder's 1785 retelling of the Echo myth dramatizes this problem, the troubling persistence of femininity despite—or through—its exclusion. Promising to tell the "true history of Echo," Herder recounts how Harmonia assisted Jupiter in his work of creation. As Jupiter created the beings of the world, "with maternal tenderness, she imparted to the newly formed being a tone, a note, which penetrates into the depths of his bosom, binds his whole existence together, and connects him with all kindred beings."[254] Yet because she is only half immortal, Harmonia is destined to die and abandon her children. She is inconsolable. In his great compassion, Jupiter grants her the ability to feel her children's wretchedness invisibly by repeating the sounds they make. Wherever a tone from one of her children is heard, Herder writes, "the heart of the mother resounds in sympathy."[255] Hence the existence of the echoes we hear today.

In his idiosyncratic etiology, Herder offers a playful alternative to the theory of language origin advanced in the *Treatise*. The role of the feminine qua maternal, however, has changed little in the fifteen years separating the two texts. Playing the mother to Jupiter's father, Harmonia "assists" and "imparts" while Jupiter acts and creates. Her "language" is pure sound: phatic communication without semantic content. As she is transformed into an echo, Herder writes, Harmonia becomes "formless, widespread."[256] What better way to describe how femininity functions in the philosophies of language examined here? Diffuse across the conceptualization of language, yet never sufficient unto themselves, feminine figures are distorted reflections of the masculinist fantasies that produce them. They act, in other words, as the masculine's echo.

"Women's Language" and the Language of Science

The texts examined in chapter 1 suggest that woman's ideal relationship to language is one of passive perpetuation. Women do not form language but nurture the inventions of men. Yet as many of those same authors would realize, often to their chagrin, women do not always recognize the roles they have been prescribed; the ideal and the real may find themselves at odds. What happens when women veer from the normative path of linguistic gestation and lactation? What happens, in other words, when women attempt to create language themselves? The result—at least according to many nineteenth-century scholars—is a monstrosity, a curiosity, a spectacle: "women's language."

In the seventeenth and early eighteenth centuries, the term *women's language* (*Weibersprache* in German, *langage des femmes* in French) is generally restricted to ethnography and ethnology. It refers to discrete languages apparently spoken by women among so-called "primitive" peoples. By the mid-1800s, however, *women's language* comes to refer not only to "primitive" women's languages but also to the general alterity of female speech. In scientific and anthropological texts about language, in particular, data on "primitive" women's languages stand alongside claims about the difference of *all* women's speech—such as Wilhelm von Humboldt's assertion in the 1820s that Weibersprache is "tied more closely to nature" and "a more faithful mirror of [women's] thoughts and feelings."[1] The "female characteristic," Humboldt writes, expresses itself in language as it does in all aspects of life.[2] Starting with seventeenth-century ethnographic accounts of a women's language in the Caribbean, then moving to nineteenth-century language science and anthropology, this chapter traces how the purported phenomenon of women's language was transformed from a contingent, "savage" occurrence into a necessary symbol of female alterity. This spurious conflation, whereby discrete women's languages are made to function as evidence for a universal theory of women's speech, helped to make possible the constitution of a scientific authority that was distinguished from a gendered as well as racialized Other. Whereas chapter 1 argued that philosophies of language theoretically justify the dehumanization of women by pulling language to man, this chapter shows how linguistic thought performed this dehumanization by associating all women with the nonwhite, the non-European, and the "primitive."

Constructing "the Primitive"

The idea of a distinct women's language first enters the European imagination in the seventeenth century, when Christian missionaries to the Caribbean write of separate men's and women's languages among the people indigenous to the Lesser Antilles. French-speaking missionaries document this phenomenon—what they term the Carib *langage des femmes*—in dictionaries and protoethnographic texts, including Mathias DuPuis's 1652 *Relation de l'establissement d'une colonie françoise dans la Gardeloupe isle de l'Amérique, et des moeurs des sauvages* (Relation of the establishment of a French colony in the American island of Guadeloupe, and the customs of the savages), Raymond Breton's 1665 *Dictionnaire caraïbe-françois* and 1666 *Dictionnaire françois-caraïbe* (Carib-French and French-Carib dictionaries), the 1667 edition of Charles de Rochefort's *Histoire naturelle et morale des Iles Antilles de l'Amérique* (Natural and moral history of America's Antilles Islands), and Jean-Baptiste Du Tertre's 1667 *Histoire générale des Antilles habitées par les François* (General history of the Antilles inhabited by the French). As these texts describe it, Island Carib women could understand and speak the language used by men, but the men either could not understand or would not dare to use the langage des femmes, for fear of ridicule. To explain this unusual situation, European settlers cite a story of mass murder and mass rape: long ago, they report, Carib men from the mainland came to one of the islands and killed the male inhabitants, enslaving the women as their wives. These women, who belonged to a different tribe, the Arawaks, continued to pass down their native tongue to their daughters, generation after generation. Some texts also cite a third language, reputed to be very difficult, spoken only by adult male warriors.[3]

Already prior to the missionaries' discussion of the langage des femmes, the "Island Carib" had been constructed as the quintessential savage by European colonizers—an "icon" of savagery, as Neil L. Whitehead puts it.[4] Beginning with the Spanish conquest in the fifteenth century, specifically a report by Christopher Columbus from 1493, "Island Caribs" were figured as brutal cannibals in contradistinction to the more peaceful tribes on the South American mainland.[5] One purpose of this "invention of the Island Carib," Whitehead observes, "was the legal and moral definition of resistant native populations in terms that permitted their unhindered conquest."[6] The seventeenth-century "discovery" of the women's language was assimilated into this characterization of the Island Caribs as ferocious savages. The langage des femmes, ostensibly born out of rape and murder, was considered further proof of the Island Caribs' brutality.

The missionaries' descriptions of the Island Caribs' origins and even the name *Island Carib* are now widely contested.[7] Scholars currently disagree about the heritage of the people who lived in the Lesser Antilles at the time of European contact and about the language(s) they spoke.[8] The subject of my study is not the history or anthropology of the Caribbean but rather the ways that "Island Caribs" have, in Whitehead's words, been made to "bear the epistemological weight of European expectation and speculation," not only licensing colonial conquest but also enabling "the definition of a European self" vis-à-vis the construction of a primitive Other.[9] The extent to which the imagination of the Island Carib as savage cannibal grounded the formation of the discipline of anthropology and the European self-definition of "civilized" has been well documented.[10] What I aim to show is that the "European self" into whose formation the "Island Caribs" were drafted was not only a racialized subject but also, by the nineteenth century, a gendered one as well. As this section details, the genealogy of European discourses on the Carib women's language reveals a significant change in the nineteenth century: a shift in emphasis from cultural to sexual difference, and from the ethnography of local peoples to a general theory of women's alterity.

In the seventeenth-century discussion of the Carib langage des femmes, notions of savagery and barbarism predominate. The first text to mention the women's language is the *Relation* by Mathias DuPuis, who served as a Dominican missionary on the island of Guadeloupe. DuPuis explains that "the women have a very different language from that of the men: & as it would be a crime for them to speak differently when they are not obliged to talk to men, so they make fun of men who use their way of speaking."[11] DuPuis does not elucidate the origins of the langage des femmes but presents it as one linguistic curiosity among many, including the fact that young and old men also have "completely different ways of speaking" and that men have yet another language that they use during war, a language so impoverished that its speakers "cannot name any virtues because they don't practice any."[12] According to his text, the Island Caribs are not only brutal in warfare but also aggressive in their everyday behaviors: "Their debauchery is frequent, not to say continual, in which they get carried away to the point of stupidity."[13] Excessive drinking, DuPuis writes, leads men to express their anger in "horribly disastrous vengeance."[14]

The langage des femmes is subsequently referenced by the Carib-French and French-Carib dictionaries of Raymond Breton. Breton, who arrived on Guadeloupe in 1635, was "one of the original four Dominican missionaries who helped establish a base for the Frères prêcheurs," his religious order, in the French West Indies.[15] Breton's French-Carib dictionary distinguishes between words used by women and men, but offers basically no commentary on this phenomenon except to note how the dictionary should be read: "The letter f. indicates that the following word is of the women's language."[16] Similarly, Breton's Carib-French dictionary mentions the women's language only in a few relevant definitions. Readers would find lengthier interpretations of the langage des femmes in the "Histories" of Jean-Baptiste Du Tertre and Charles de Rochefort, who were Catholic and Protestant missionaries, respectively.[17]

Compared to earlier Spanish reports of the Island Caribs as vicious cannibals,[18] Du Tertre's Histoire générale and Rochefort's Histoire naturelle et morale offer a more sympathetic view of the people indigenous to the Lesser Antilles, although they still trade on the idea that the Island Caribs are more vicious than the people on the mainland. Du Tertre's text in particular attempts to balance two potentially opposing views of the Island Caribs. On the one hand, he portrays them as noble savages: "They are as nature made them," Du Tertre writes, "that is to say, with a great simplicity and natural naivete."[19] On the other hand, Du Tertre, in the words of historian Philip P. Boucher, "accepts the orthodox Christian view that natural man tends towards evil," which leads him to condemn many of the practices he encounters.[20] Regardless, however, of whether Du Tertre is portraying the Caribs as noble savages or as deplorable heathens, the relationship he constructs between Caribs and Christian Europeans is seemingly always based on difference. "I have yet to speak of the birds in the air, the fishes in the sea, rivers & ponds," he writes in the introduction to the second volume of his Histoire, "& to make known their difference with Europe's birds and fishes."[21] This method—contrast with Europe, rather than comparison—is Du Tertre's method for describing not only the birds and fish of the islands but the people as well.

Although first published in 1654, it is not until the revised 1667 edition, after the publication of Breton's dictionary, that Du Tertre's Histoire makes any mention of a langage des femmes. Even then, the existence of men's and women's languages is referenced only in passing, as further evidence for a certain interpretation of, as the chapter title puts it, "The origin of our islands' savages." Du Tertre asserts that the Island Caribs are related to the mainland

Caribs ("Galibis") and offers several reasons to support this argument. First, there is "the common custom of all Savages"; second

> is the diversity in the men and women's language, which still remains today; because they say that this diversity originated at the time of that conquest, especially given that Galibis had killed all the males of those Islands, & saved only women & girls to whom they gave young men of their nation to marry, each kept their original language. If you consider furthermore the conformity in religion, customs & language, there is no reason to doubt that they originate from the Galibi people from terra firma.[22]

For Du Tertre, the existence of the langage des femmes corroborates the understanding of the Island Caribs as brutal and ferocious. Notably, however, this is because it testifies to the murder of the men, not the rape and sexual enslavement of women. The mass rape that this event presumably entailed is glossed over as "marriage" and not discussed further. Here, Du Tertre is in line with his contemporaries, who similarly call the rape and abduction of women simply "taking them as wives." The murder of the men on the island, on the other hand, receives more attention and its brutality is described in detail. Du Tertre, for instance, documents the Caribs' custom of decapitating and keeping the heads of enemies. Quoting from Breton's definition of Galibi, Du Tertre explains that the leader of the Caribs, who left the mainland to conquer the islands, killed the original inhabitants,

> with the exception of the women, who always kept something of their language, that to keep the memory of his conquests, he (the captain) had the heads of his enemies (that the French have found) brought to the rocky dens by the seashore, so the fathers could show them to their children, and after them all those who would descend from them.[23]

One might draw a corollary between the Caribs' conquest of the islands and European colonization, but the mass extermination of men marks, in the ideology of Du Tertre's text, a fundamental different between the two. Even though Du Tertre criticizes European colonizers for specific incidents of violence against the Caribs, his text operates with the assumption that the French have come to the West Indies in the name of civilization and Christianity—and that their possession of the islands is supposed to be of a different order than the Carib capture of the Arawaks. For all his sympathy for the Caribs' noble, "primitive" way of life, Du Tertre does not question the general colonial project.

Du Tertre's Huguenot contemporary Charles de Rochefort provides basically the same account of the origin of the Carib women's language, although he also elaborates, in a way that Du Tertre does not, on the more practical aspect of how the langage des femmes was perpetuated:

> The Savages of Dominica say that this proceeds from the fact that when the Caribs came to settle the Islands, they were already occupied by the Arawak Nation, which they destroyed entirely, with the exception of the women, whom they married to populate the country. So that these women, having kept their language, taught it to their daughters, and got them accustomed to speak it like them. This having been implemented until now by Mothers with daughters, this Language thus remained different than that of men in many aspects. But the boys, although they understand their mothers' and sisters' tongue, nevertheless follow their Fathers and brothers, and fashion them-selves to their Language, as soon as the age of five or six.[24]

In Rochefort's formulation, the Caribs women's language persists because the sons reject their mother tongue. Whether these sons will ultimately forget their mothers' language, or whether they will always retain the capacity to understand it, Rochefort does not say. But it is noteworthy that in Roche-fort's text, the continuation of the women's language has more to do with the agency of men than of women. It is the fathers and sons who keep the women's language at bay, who prevent it from becoming dominant. The lan-gage des femmes is in no way a secret language guarded by women, as it will be interpreted by later texts.

It is important not to overstate the significance that Rochefort affords to the women's language. He devotes only a small portion of the chapter on lan-guage to discussing the langage des femmes, as it is just one of a number of examples of what he understands to be the particularity of the people he has encountered. Rochefort practices a kind of relativism—"the Caribs have an ancient and natural language, which is specific to them, just like every Nation has its own"[25]—in which a people's language is a reflection of their distinct customs. Consequently, the text documents numerous traits that distinguish the Carib language from those spoken in Europe. First, there are all the things it apparently lacks: the letter *p*,[26] words for *winter* or *ice*, any number higher than ten, comparatives and superlatives, a word for *soul*, more than four col-ors.[27] Then there are the "features of their language's naivete and elegance," like the fact that Caribs apparently call lips "the edge of the mouth."[28] The langage des femmes is not granted a special revelatory status but is rather one of many distinctive features of the Caribs that Rochefort's text presents for its readers—including, according to the list in his opening epistle, "the obscurity of their origin, the harshness of their language, the Barbarism of their customs, their odd way of living, the cruelty of their wars, their ancient poverty."[29]

In the early 1700s, the account invoked by Du Tertre and Rochefort remains the dominant explanation of the langage des femmes. What changes during this period, however, is that instead of being considered exclusive to the Island Caribs, women's language is now conceptualized as a general phenomenon of "savage" peoples. Joseph-François Lafitau, a Jesuit missionary and naturalist, turns to Herodotus to find further evidence for the theory that the abduction and sexual enslavement of women can lead to the development of a separate women's language. Although Lafitau did not encounter the Island Caribs directly, he studied earlier accounts of the Caribs and other native peoples, in addition to spending six years working as a missionary among the Iroquois.[30]

"Herodotus tells of a very strange fact, from which one could shed some light on the origin of the Caribs of the Antillean Islands," Lafitau suggests in his 1724 study, *Moeurs des sauvages amériquains comparées aux moeurs des premiers temps* (Customs of the American savages compared to the customs of the first ages). Here, Lafitau recalls Herodotus's story of a group of Ionians that murdered Carian men and took their women as "wives." Because of this massacre, Lafitau writes, the captured women

> took an oath to never eat with their husbands, and to never speak their name; and made it a Law to transmit this custom to their posterity, by teaching the children who would be born of these unions: that this custom was because the Victors had slit the throats of their fathers, husbands and children. Carib women thus never eat with their husbands; they never call them by their name; they serve them as if they were their slaves: and what is even more odd, is that they have an entirely different language than that of their husbands, as probably Carian women had, who were foreign to these people who came from Euboea and brought desolation to their country. One could add that there is a link between the Carian and Carib names, which is used nowadays by the Savages I am talking about.[31]

Whereas seventeenth-century writers characterize the women's language as a relatively organic occurrence (the Arawak women spoke a different language and thus continued to speak it), Lafitau turns it into a conscious *choice*, a means of revenge. Refusing to call their husbands by name is, of course, different from speaking an altogether distinct language; Lafitau nonetheless conflates the two in one sweeping sentence. In his interpretation, the survival of the langage des femmes is the women's retaliation for the enslavement and murder of their ancestors.

By turning to Herodotus, Lafitau appears to naturalize the women's language of the Caribbean: the phenomenon is recognizable according to recorded traditions.[32] Lafitau's idiosyncratic etymology even emphasizes this
link, drawing a connection between the names *Carib* and *Carian*.[33] In Lafitau's
description, we can also see a twisted reformulation of the ancient Philomela myth, which similarly thematizes sexual violence and linguistic punishment. When she is brutally raped by her sister's husband, Tereus, Philomela—
according to Ovid's account—"voic[es] her sense of outrage and cri[es] her
father's name."[34] Tereus cuts out her tongue to keep her silent, but Philomela
weaves her story into cloth and thereby communicates with her sister, Procne.
The sisters take their revenge on the debauched Tereus by killing and feeding
him Procne and Tereus's own son; the story ends as the women are transformed into birds. While Philomela's silence is brutally imposed, the Carib
women supposedly choose a vow of silence. And while the Carib women perpetuate this silence through their children, Procne destroys the possibility of
future generations altogether. Yet despite these differences, there are striking
parallels in how the stories constellate the themes of rape, silence, murder, and
revenge.

These parallels are, to a certain extent, the point. Comparing the "savages" to the "ancients" is the stated objective of Lafitau's text, part of what has
been called his "theory of global culture" and methodology of "reciprocal il-
lumination."[35] Because he wants to argue against atheists who claim that "natural man" has no inclination toward religion, he must contend that "the
striking similarities between the religious traditions of the Chinese, the American savage, the antique Greek and Hebrew, and the Christian could be sat-
isfactorily explained only by the hypothesis that true religion existed long
before Moses."[36] Yet this argument still insists on an essential gap between
Caribs and modern Europeans, since Lafitau must look to Europe's *ancient*
past—to the barbaric prehistory of Greek civilization—in order to find the
Carians as an analogue to the Caribs. As Sara Petrella observes in her study of
the illustrations of "savage" women in Lafitau's text, the *Moeurs des sauvages*
exemplifies an intermediary period between the sixteenth century's conceptualization of the savage as monstrous Other and the biologized notions of
alterity of the nineteenth century.[37] This is borne out in Lafitau's analysis of
the langage des femmes: rather than a phenomenon distinct to the Antilles, in
Lafitau's text, women's language is a custom found among "uncivilized" peoples throughout history.

The notion that the existence of distinct men's and women's languages
signifies a people's violence and primitivism gains further traction over the

course of the eighteenth century. In 1738, the Englishman Francis Moore issues a description of life along the River Gambia. Moore had worked as a clerk for the Royal African Company, an institution that for almost seventy years had viciously enslaved African people and shipped them across the Atlantic.[38] Moore's experience in Africa as a clerk provided fodder for his *Travels into the Inland Parts of Africa*, where he describes a group of people who have a language "entirely unknown to the Women, being only spoken by the Men," which is "seldom us'd by them in any other Discourse than concerning a dreadful Bugbear to the Women, call'd *Mumbo-Jumbo*, which is what keeps the Women in Awe: And tho' they should chance to understand this Language, yet were the men to know it, they would certainly murder them."[39] It is this purported "men's language" that, fifty years later, secures the attention of Christoph Meiners in his *Geschichte des weiblichen Geschlechts* (History of the female sex). Meiners was a professor of philosophy in Göttingen; he was also "cultural racist" who constructed a hierarchy of the world's *Völker*.[40] As Caroline Franklin has demonstrated, Meiners was not a particularly original thinker in his attitudes on sex and race, but rather a "mediator, synthesizer and reformulator of Enlightenment thought for [his] middle-class readers."[41]

In the *Geschichte des weiblichen Geschlechts*, Meiners argues that a people's level of development can be measured by its treatment of women and by its women's sexuality.[42] Writing about the same African tribe as Moore, Meiners goes beyond Moore's description in that he compares the men's language of the "Mandingos" to that of the Caribs, and thereby turns the existence of discrete men's and women's languages into a trait common among "savage" peoples, or—as he puts it—*Wilden*. (In reference to the Caribs, this is the language of adult male warriors mentioned by Rochefort and others.) For Meiners, the violence that these men use to keep women from learning their language is supposed to illustrate the baseness of their society:

> The overbearing power that they wield over their wives is not enough; instead, all men have established a terrible alliance, and a secret teaching against women, which the Spanish Inquisition has nothing on. First of all, *like the Caribs*, they have a secret language, which the men only speak among themselves when they want to do something to the women. If a woman were curious enough to learn this language, as soon as the men got wind of her learning, she would certainly be slain without mercy as a betrayer and defiler of sacred secrets. No less impenetrable is a secret order of men that was founded to tame or punish guilty women. No youth under sixteen is accepted into this secret alliance and whoever is initiated into it must swear the most terrible oath that he will not reveal the secrets of the order to the uninitiated, and least

of all to a person of the opposite sex. The violation of this vow is punishable
by a death penalty that cannot be excused even for the most distinguished
persons.[43]

Because he is arguing for a hierarchy of the world's nations—based, no less,
on the status of the women in them—Meiners's description here is implicitly
contrasted with his description of women among "higher" peoples, where
there is ostensibly no division between the languages of the sexes. In Meiners's text, the existence of separate men's and women's languages is coterminous with unrestrained violence, and thus with "uncivilized" society.

The German explorer and naturalist Alexander von Humboldt also discusses the Carib women's language in his "personal narrative," the three-
volume *Relation historique du Voyage aux Régions équinoxiales du Nouveau
Continent* (Historical relation of the voyage to the equatorial regions of the
new continent). This text was the result of the notes and diaries he kept during a voyage to South America beginning in 1799. Humboldt is significant
insofar as his text stands as a bridge between the perspective on women's language found in the 1700s (women's language as a practice of the uncivilized,
located at a great historical or cultural distance) and the "among us, too" interpretation that becomes prevalent in the century that follows. Because he
is a German who writes his travelogue in French, Alexander von Humboldt
also acts as an intermediary between the discussion of the Carib women's
language in French ethnographic discourse and the interpretation of women's
language in German linguistic thought, particularly that of his brother, Wilhelm von Humboldt.

In his study of the women's language of the Lesser Antilles, Alexander
von Humboldt vacillates between two perspectives. On the one hand, his text
maintains that this phenomenon, women speaking differently, is particular
to and characteristic of the tribes of the Americas. Humboldt argues that
sex-based language differences are also to be found, albeit to a lesser degree,
"among other American nations (Omagua, Guarani, Chicquito), in which it
only diverges in a small number of ideas, for example, in the words for mother
and child."[44] On the other hand, Humboldt then suggests a more universal explanation. "Cicero had already noticed that ancient forms remained mostly
in women's mouths," he writes, "because their position in society makes them
less exposed to these vicissitudes of life (to these changes of place and occupation) that, for men, tend to alter the language's primitive purity."[45]

Humboldt ultimately dismisses this dictum as insufficient for explaining
the women's language of the Caribs, which is too "grand" and "surprising" to
simply be an example of what Cicero describes. Instead, he chooses the *rap-*

tio theory favored by the French missionaries, even emphasizing the violent nature of the Carib takeover—"they arrived there as a pack of warriors, not as settlers accompanied by their families."[46] Nonetheless, the mere fact that Humboldt suggests Cicero's explanation as a possibility is noteworthy, especially when we compare it to the prior 150 years of writing on the langage des femmes, which made no connection between the women's language of the Caribs and general qualities of women's speech. This is not Lafitau's equation of American "savages" with ancient barbarians; quite to the contrary, in the passage that Humboldt cites from Cicero, the speaker Crassus refers to the language of his mother-in-law, a member of the Roman upper class. Whereas in Lafitau, women's language was a more general, primitive feature within an overall history of the human species' progress, in the case of Alexander von Humboldt we have a universalizing take on women's language that is not bound up with women's lack of access to a specific culture, but rather with the possibility of a distinctly *female* linguistic practice.

Humboldt's *Relation historique* is a hybrid text in which, as Oliver Lubrich argues, many voices and registers compete for authority.[47] Humboldt's discussion of women's language is indicative of the general tension that defines his text. Is the Carib langage des femmes a "savage" custom, found only among the "wild" people of the Americas, or is it a widespread, "female" phenomenon—found even among the Roman aristocracy? How much, in other words, do "civilized" Europeans and "primitive natives" have in common? And which is the ultimate divisor: cultural or sexual difference? Humboldt's text cannot quite decide. And in this ambivalence, he represents a new interpretation of the langage des femmes, in which woman stands as a link between civilization and savagery.

Historical Characterizations of Women's Speech

When missionaries introduced the Carib langage des femmes to Europe in the seventeenth century, general ideas about women's speech were of course already in existence. The notion that women speak differently than men, and that this difference extends beyond timbre and pitch to the style, content, and *quality* of women's speech, can be found already in antiquity. I am interested in how nineteenth-century texts on language take up conventional ideas about the difference of female speech, setting them alongside discussions of "primitive" women's languages around the world. The convergence of these two discourses, the popular and the ethnological, is an important moment of transformation. It is at their intersection that women's language—formerly only a "strange" characteristic of the Caribs—is actively theorized, crystallized into a concept that means *both* a discrete women's language *and* a universal female way of speaking. The following section provides a brief account of the different ways that women's speech has been characterized historically so that we may see how they inform later "scientific" discussions of women's language.

Ancient characterizations of women's speech cohere around several tropes: women's speech as excessive, as nonsensical, and as weak. Among the many distinctions between men and women documented by Aristotle's *History of Animals,* for example, is that women are more "querulous" and more "false of speech."[48] Similarly, the *Physiognomics* of the Aristotelian school, which claims that the male body is "better conditioned and more fit for every function" than the female, correlates the female voice with morbidity.[49] The Aristotelian *Problems* likewise argues that women, like other "persons without generative power, such as boys [. . .], men grown old, and eunuchs," have shrill voices because their breathing is "feeble and sets little air in motion."[50] The Roman rhetorician Quintilian makes a similar assertion when he argues, in Thorsten Fögen's paraphrase, that "a feeble and thin voice is associated with female speech and thus by all means to be avoided by the future orator."[51] Anne Carson has also shown how the female voice was negatively classified in antiquity, associated with a dangerous sexuality that threatened the polis and had to be rigorously contained.[52]

In the ancient imagination, women are not creators of language but vehicles for its preservation. In the *Cratylus,* Plato has Socrates argue that women are not as good at creating names as men, while also describing women as being the best "preservers of the ancient language."[53] And as we saw in the previous section on Alexander von Humboldt, Cicero makes an analogous claim, arguing that it is easier for women to "keep the purity of antiquity, because, by keeping less company than men, they always stick to what they first learned."[54]

Another trope popular in antiquity was women's volubility—the notion that women speak relentlessly, incessantly, and on topics of little substance. This appears in Semonides's "Types of Women," which catalogs women according to the materials that Zeus supposedly used to create them. One especially bad type, made from "a bitch, ill-tempered," is the sort of woman who never stops talking: "A man can't stop her with threats, nor even if in anger he should knock out her teeth with a stone, nor can he by speaking to her soothingly, not even if she happens to be sitting among guests, but she constantly keeps up her yapping which nothing can be done about."[55] Of the ten types of women that Semonides documents, only one is favorable: the woman made from the bee, who takes no pleasure in gossip, or in "sitting among women in places where they talk about sex."[56] Juvenal's sixth satire, written between the first and second century AD, similarly catalogs the ills of excessive female speech. Here, women's loquaciousness is linked to other immoderate behaviors. Even more intolerable than a woman's excessive eating and drinking is her attempt to speak on intellectual matters: "So thick, and fast, / The wordy shower descends, that you would swear, / A thousand bells were jangling in your ear, / A thousand basins clattering."[57]

Disparaging characterizations of women's speech are also found in medieval and early modern Europe, where women in particular were considered susceptible to "bad" and excessive speech acts. The crime of being a "scold," for instance, a punishable offense in medieval and early modern England, was associated with femininity, and the majority of accused offenders were women.[58] The "sins of the tongue" were similarly feminized transgressions.[59] And as Emily Butterworth shows in her study of Renaissance France, *caquet* (babble or excessive talk) "was and is a gendered term in French, suggesting a notion of unofficial, unregulated female chatter."[60] The German analog of caquet is *Weibergeschwätz* (women's gossip, babble, prattle). In the German-speaking lands of medieval and early modern Europe, Weibergeschwätz was negatively contrasted with *Schweigsamkeit* (reticence or quietness),[61] a principle bolstered by the valorization of women's silence in Christian doctrine.[62] In popular writings of the period, Weibergeschwätz is associated with the places where women assemble, such as the spinning wheel, the town well, the laundry, and even the road—a satirical text from 1512, for example, complains that women gather on the streets to talk instead of going to church or staying at home.[63] The derogatory connotations of Weibergeschwätz are borne out in a myriad of proverbs and aphorisms, many of which have been collected by Pia Holenstein and Norbert Schindler. They include: "Es ist beim Weib verschlossen / Wie Wasser in ein Sieb gegossen" (A woman keeps a secret like a sieve holds water); "Weiber findet man nimmer ohne Rede" (Women are never found not speaking); "Für die Weiber ist Schweigen/Härter als Säugen" (For women, silence is harder than nursing); "Die Weiber führen das Schwert im Maul, darum muß man sie auf die Scheide schlagen" (Women carry a sword in their maw, that's why you have to hit them on the sheath [vagina]); and "Kein Kleid steht eine Frau besser als Schweigen" (No dress looks better on a woman than silence).[64]

When German language purists campaigned against the use of Latin and French words in the seventeenth century, they frequently blamed women for degrading German and classified women as *Sprachverderber* (language corruptors).[65] According to William Jervis Jones, women who used Latin instead of German terms were accused of having loose morals; it was assumed that they could only have learned Latin by sleeping with male students.[66] The connection between sexual and linguistic chastity was a common strategy of the period. A 1640 lamentation by the poet and scholar Justus Georg Schottelius, for example, personifies the German language as a woman who has fallen into prostitution because she has been debased by foreign terms.[67]

What *causes* women's loquacity, their tendency to degrade language through excessive talk? To construct etiologies of women's volubility, writers have appealed to a variety of authorities, including God, physiology, and nature. Nicolas de Cholières's 1587 *Après Disnées* attributes "the female tendency to chatter to the moistness of their brains,"[68] as Robert Muchembled accounts. "In the end, chatter is of great use" to women, de Cholières writes, "purging the brain and evacuating the bad humours that could at length, if held in, cause them ill."[69] In the following century, a different French text explains that "since God made the body of woman from a hard, creaking rib, and that of man from the deaf and dumb earth, it was a foregone conclusion that man would be by his nature taciturn and silent."[70] Eighteenth-century English critics constructed different explanations for women's excessive talk. Patricia Howell Michaelson documents several choice examples, including those of Joseph Addison, who speculated "that there are certain Juices which render a woman's tongue so voluble and flippant,"[71] and William Smellie, who derived "women's quantity of speech from the needs of child rearing, women's natural role."[72] According to Giambattista Vico, on the other hand, women—like the "feeble-minded"—are naturally prone to digression because they are unable to focus on what is essential.[73]

In the eighteenth century, etiquette books and advice manuals, which include sections on sex-specific speech norms, perpetuate the idea that women speak best when they speak meekly. Joachim Heinrich Campe's *Väterlicher Rath für meine Tochter* (Fatherly advice to my daughter), published in 1789 and reissued several times hence, advises a young woman that, if she must disagree with someone, take care "that your face is always friendly, your voice gentle, your opposition itself modest and shy."[74] Similarly, *The Ladies' Book of Etiquette and Manual of Politeness*, first published in Boston in 1860, includes two different chapters—on conversation and on letter-writing—that discuss women's language use. Concerned as much with class as with gender norms, *The Ladies' Book* warns against speaking authoritatively, using or recognizing double entendres, and other practices that could indicate "ill-breeding."[75] Prescriptive statements about women's speech were not limited to those aspiring to upward class mobility, however. The handbook *Das häusliche Glück. Vollständiger Haushaltungsunterricht nebst Anleitung zum Kochen für Arbeiterfrauen* (Domestic Happiness. Complete Instruction in Homemaking Including Instructions for Cooking for Working-Class Women), first published in 1882, includes a section that instructs women to "keep quietly to yourself, as far away as possible from gossiping female friends."[76]

Although they differ in terms of the specific practices they advocate, these etiquette books agree that women's speech should be curbed and contained. Indeed, when woman is even regarded as a potential orator—and this doesn't happen seriously until the end of the eighteenth century—she is considered to be at her best when both her speech, and the context of its utterance, operate under a principle of restraint. In the Renaissance, women were prohibited from learning rhetoric, as Patricia Parker shows.[77] In the late eighteenth century, when, according to Lily Tonger-Erk, women's rhetoric became a topic of discussion, while oratorial norms for men presupposed the context of male speech to be the public sphere, women's rhetoric was considered valuable only in the private sphere, in her conversations with her family and, especially, her husband.[78] This valorization of female silence and verbal restraint reverberates throughout the nineteenth century. The fairy tales of the Brothers Grimm, for example, correlate female speech with slothfulness, as Ruth Bottigheimer has argued.[79]

A turn to historical semantics can help elucidate how various derogatory characterizations of women's speech coalesce into a codified term in nineteenth-century Germany. Significantly, the term *women's language* has a more extensive history in German than in French. While the phrase *langage des femmes* appears in print as early as the mid-sixteenth century, in a translation of Herodotus,[80] it refers there to a specific group (the language *of the Amazon women*) rather than a generalized concept (the language *of women*). Moreover, although early French dictionaries do, at times, discuss *voix féminine* (feminine voice) under the entry for *Féminin* (feminine), they define it only as "resembling [that of] a woman." Such dictionaries include no entries for *langage des femmes* or other equivalent terms.[81] Likewise, the section on "Langage" in Diderot and d'Alembert's eighteenth-century encyclopedia makes a case for relativity among nations—arguing that culture and climate influence the vocabulary and expression of a people's language—but there is no discussion of differences in language use between the sexes. While there were of course stereotypes about women's speech in existence during this period, the term *langage des femmes* appears to have had no referent in French before its introduction in the ethnographic context. English similarly has little trace of *women's language* as a generalized term prior to its use in ethnographic texts.[82]

By the end of the seventeenth century, however, *Weibersprache* can already be found in the German lexicon as a derogatory term whose meaning is unrelated to the women's language documented in the Antilles. And even earlier than that, German dictionaries are recording terms for women's poor language use. Josua Maaler's 1561 dictionary *Die Teütsch spraach* (The German language) includes an entry for *weybische/unmannliche außsprächung* (womanish/unmanly pronunciation), which he defines as "fracta pronunciatio" (broken pronunciation).[83] In the 1691 dictionary *Der teutschen Sprache Stammbaum und Fortwachs, oder Teutscher Sprachschatz* (The family tree and development of the German language, or German lexicon), Kaspar von Stieler lists *Weibersprache* as its own term, defining "jungfer- und Weibersprache" as "sermo tinniens, vox acuta" (shrill speech, a sharp voice).[84] Stieler's definition of *Weibersprache* resurfaces in the Grimms' *Deutsches Wörterbuch* (German dictionary), which also adds a second reference to the Caribs.[85] The entry in Stieler for *men's language*, on the other hand, is much more positive: "sermo robustus, oratio mascula, valida, fortis, gravis" (firm speech, language that is masculine, effective, strong, serious).[86] In Stieler's dictionary, *Weibersprache* is furthermore differentiated from *Weiberstimme*, which receives its own entry as "vox foeminea, exilis, tinniens" (a voice that is feminine, feeble, shrill),[87] and which would be analogous to the French *voix féminine*.

Charles de Rochefort's *Histoire naturelle et morale*, one of the earliest texts to report on the Carib women's language, was translated into German in 1668, but this edition remained relatively obscure, significantly less well known than the popular English translation.[88] Du Tertre's *Histoire générale* was never translated into German; the same is true for another seventeenth-century French text that would have informed its readers of the Carib women's language, Antoine Biet's 1664 *Voyage de la France équinoxiale* (Voyage to equatorial France). Stieler's "shrill speech" definition of *Weibersprache* thus predates the widespread introduction into German of French ethnographic writing on the Caribbean. This may help us understand why it was predominantly German texts, rather than French or English ones, that initiated the practice of conflating the Carib women's language with a general theory of women's speech. German already had a codified term for popular ideas about women's speech, while French and English did not.

Of course, a coincidence of terminology does not a new concept make. The conflation of *women's languages* with *women's speech* would not have been possible without a reconceptualization of language and its study—without the emergence of disciplines that necessitated empirical data and the purification of their own language into "precise" discourse. In the 1800s, a period in which *Sprachwissenschaftler* were striving to establish the scientificity of their discipline and ground their authority,[89] "women's language" becomes a productive foil. Both in the way it is cataloged and verified, and in the way it is constructed in opposition to the language *of* language science, the fascination with Weibersprache is best understood as a symptom of this discipline's anxieties and a product of its efforts at self-legitimation.

Weibersprache Universalized

It is Johann Gottfried Herder's *Treatise on the Origin of Language*, which I examined in chapter 1, that brings the Carib Weibersprache into the German discourse on language. While Herder does not analyze the women's language in detail, he raises the possibility that women's difference expresses itself in a distinct female culture, which informs later conceptions of "women's language" in philosophy, anthropology, and language science.

For Herder, Weibersprache is an extreme example of the linguistic superfluity that marks all languages, which is incommensurate with divine order. Having men and women speak two different languages doubles a people's vocabulary, creating an "excess of synonyms."[90] Herder argues that this kind of linguistic excess is best explained by a human origin. In the early days of humanity, people lived nomadically and in separate groups. Then "when they afterwards came together, when their languages flowed into an ocean of vocabulary, the more synonyms!"[91] Herder does not provide an etiology of the Carib Weibersprache, but his normative concept of the *Familienfortbildung der Sprache*—in which it is the father who originates language, teaching it to his wife and child—would support the explanations offered by the missionaries' texts. Carib Weibersprache is not a language that women *invented*, but simply one that they perpetuated as objects of exchange between men.

Another brief discussion in the *Treatise* confirms this interpretation: women's languages are portrayed as the consequence of unenlightened customs. What a *"treasure familial language is* for a developing race!" Herder writes, before outlining the significance of fathers' ballads as "the origin of the tribal language" for peoples around the world. "And it is only precisely here that it can be indicated," he continues,

> why among so many peoples, of which we have cited examples, the *male* and *female* sexes have almost two different languages, namely, because in accordance with the customs of the nation the two, as the noble and base sexes, almost constitute two quite separate peoples, who do not even eat together. According, then, to whether the upbringing was paternal or maternal, the language too inevitably becomes either *father-* or *mother-tongue*—as, in accordance with the customs of the Romans, it even became lingua vernacula.[92]

Insofar as he describes women as functioning as a "separate people," Herder—
whose philosophy asserts an inextricable connection between language and
Volk—would seem to suggest something like a discrete women's culture.
Yet Herder is clear that such instances do not result from women's self-
determination or expression, but from a patriarchal culture that excludes
them as "base." Whereas Herder's *Treatise* has little investment in discussing
women's languages, nineteenth-century authors would engage more thoroughly with the topic, offering explanations that present Weibersprache as a
universal phenomenon.

Wilhelm von Humboldt is one such author. In "Ueber die Verschiedenheiten des menschlichen Sprachbaues" ("On the Diversity of Human Language Structure"), written between 1827 and 1829, Humboldt devotes a sub-
stantial section to the theorization and investigation of Weibersprache. "Ueber
die Verschiedenheiten" is an early version of Humboldt's well-known *Kawi-
Einleitung*, which I discussed in chapter 1. Both versions of the text adhere to
a fundamental tenet of Humboldt's language philosophy, namely that every
language embodies a distinctive worldview, which reciprocally structures human cognition. Or as Humboldt himself puts it, languages are not "actually a
means [. . .] of representing truth that has already been known, but far more
of discovering what was previously unknown. Their difference is not one of
sounds and signs, but a difference of worldviews themselves."[93]

In the 1820 lecture "Ueber das vergleichende Sprachstudium" (On comparative language study), Humboldt makes a forceful argument for his subject. The investigation of language should not be subordinated to other fields,
he contends, but made into a science that "supports its usefulness and purpose in itself."[94] In order to establish this discipline—what he variously calls
Sprachstudium, *Sprachkunde*, and *Sprachwissenschaft*—Humboldt borrows
from the language of the natural sciences. Using terms like *dissection* (*Zer-
gliederung*) and *measurement* (*Ausmessung*),[95] he insists that the study of language must strive for "data" and "strict factual verification."[96] According to
Humboldt, languages are organisms analogous to the biological organisms
studied by the natural sciences. As he calls for an "investigation into the organism of languages,"[97] he repeatedly asserts that language is "an organic being and must be treated as such."[98] This means examining languages empirically and systematically. "Ueber das vergleichende Sprachstudium" outlines
a general system of comparative investigation, including the initial step of
studying a language's internal coherence.[99] Humboldt employs this kind of
systematic approach in his study of the Basque language, which he publishes
the following year.[100]

For Humboldt and his contemporaries, language science is not only *comparable* to the natural sciences but actually exceeds them in its capabilities, since it is language that defines the human at his most essential—"the human is only human through language."[101] While grammar may have "its anatomy, its physiology, its chemical analysis as good as the rest," as a reviewer of Jacob Grimm contends in an 1822 text,[102] Sprachwissenschaft is not just one science among many but a unique discipline that can reveal fundamental truths about human history and progress.[103]As Sarah Pourciau writes, summarizing Grimm's approach, the job of the linguist is to use scientific analysis to make language "'speak' its own interior structure," revealing the spirit that forms not just language, but the world.[104]

If the ideal language *of* language science is a language of precision, women's language is its opposite—at least according to Humboldt's characterization. In "Ueber die Verschiedenheiten," Humboldt explains that the differences of sex and age are "given by nature" and necessarily identifiable in language. In particular, the "female characteristic" (*weibliche Eigenthümlichkeit*),

> which is expressed in so lively and visible a manner in intellectual matters, also naturally extends to language. As a rule, women express themselves more naturally, more tenderly and nonetheless more forcefully than men. Their speech is a more faithful mirror of their thoughts and feelings, and even though this has not often been recognized and said, they excellently retain language's fullness, strength, and accordance with nature in the midst of this culture, in which these properties are always robbed, by departing alongside men at the same pace. They thereby reduce the disadvantage of the split that culture always creates between the people and what's left of the nation.[105]

Humboldt's description of women's language aligns with the conceptualization of femininity as passive materiality found in his *Horen* essays. Similar to how femininity functions in the creation of the beautiful, women play a retarding role in the creation of language, slowing the effect of men's cultural transformations. Humboldt may describe this retarding function positively, but it is worth noting that it affords women no agency. Women are channels through which language is maintained: they "*preserve* language's fullness, strength, and accordance with nature." In what follows, Humboldt elaborates even further on the relationship between women, nature, and language. "Truly tied more closely to nature through their being," he writes,

> placed by the most important and yet the most ordinary events of their life in greater likeness with their whole sex, occupied in a way that makes use of the most natural feelings, or which grants the inner life of thoughts and

sensibilities complete leisure, free from everything that imprints upon the
mind a one-sided form like the life of business and even science [*Wissenschaft*], quite often in a conflict between outer constraint and inner desire that
even though painful, nevertheless reacts fruitfully to the disposition, often
in need of persuasion and inclined to speech through inner liveliness and
activity, women refine and beautify language's conformity with nature [*die
Naturgemässheit der Sprache*], without robbing it, or injuring it. Their influence passes so imperceptibly from family life into daily dealings that it cannot
be isolated in detail.[106]

Humboldt seems to be channeling Rousseau here, who makes a similar claim
in *Émile*: "The male is male only at certain moments. The female is female her
whole life or at least during her whole youth. Everything constantly recalls
her sex to her; and, to fulfill its functions well, she needs a constitution which
corresponds to it."[107] When Humboldt writes that women "refine and beautify
language's conformity with nature, without robbing it, or injuring it," he is not
only claiming that women don't *harm* language, but also that they don't form
it: their speech has no influence on a language that has already been created.

In Humboldt's portrayal, women's language is positively marked by a lack
of systematicity (no "one-sided forms," thoughts and feelings have "complete
leisure"). This also means that women's language is not rigorous. Indeed,
Humboldt makes this explicit when he puts a new twist on an old idea about
women's speech. That women lead lives sheltered from the world of commerce was Cicero's way of accounting for women's preservation of old linguistic forms; we saw Alexander von Humboldt employ a similar explanation in
his interpretation of the langage des femmes of the Americas. In the passage
cited above, Wilhelm von Humboldt also calls up this trope, but expands it,
shutting out women not only from business, but also science (*Wissenschaft*).

What exactly *comprises* the difference of women's language? Humboldt
does not answer this question, writing only that it "hasn't often been recognized" and "can't be isolated in detail." Humboldt's vague language continues
in the section that follows. "The female characteristic," Humboldt writes a second time,

> does not bring forth, in the way just mentioned, a distinct language, rather
> only a distinct spirit in the handling of the common one. Even upon closer in-
> spection, hardly any individual expressions and phrases would be discovered
> that would be more specific to the other sex than to ours. Cicero, meanwhile,
> testifies from his experience that outdated expressions are preserved longer
> in the mouths of women, which, since the same is found among our people,
> confirms what was said earlier.[108]

As the "ours" demonstrates, Humboldt asserts an irreconcilable divide between the language of women and that of his own text, constructing scientific discourse in contradistinction to women's speech. As Catherine Hobbs Peaden observes, the feminization of nonscientific language was a strategy employed by John Locke, Étienne Bonnot de Condillac, and Francis Bacon, among others.[109] In the nineteenth century, Humboldt's contemporaries in the discipline of history also engaged in what Falko Schnicke calls the "production of anti-scientific femininity."[110] Yet despite this robust tradition of masculinizing scientific discourse, and despite the stark divide that Humboldt draws between himself and his subject, his descriptions conform remarkably to the kind of vague and imprecise language that he ascribes to the opposite sex. What, precisely, does "a distinct spirit in the handling of a common language" mean? If "exact attention" would only discover a few phrases of distinction between women and men, then why devote such space to this topic? The difference of women's language use, presented as incontrovertibly true yet impossible to isolate and record, poses a challenge to the discipline of Sprachwissenschaft insofar as language science requires, by Humboldt's own standards, exact empirical measurements and "strict factual verification."

It also poses a challenge to Humboldt's self-professed commitment to clarity. Regrettably, the style of Humboldt's early publications had not been well received. As Paul Robinson Sweet recounts in his biography, several of Humboldt's texts from the 1790s, including the *Horen* essays, were criticized by his contemporaries for being too opaque.[111] Kant famously remarked that "he could not figure out what Humboldt had been trying to say,"[112] while Schiller promised his publisher that his journal "would not contain any more of the difficult, 'obscurely written articles such as those by Humboldt are said to be.'"[113]

If Humboldt's opening remarks about Weibersprache are vague and imprecise—full of abstract terms, lacking in examples, the only reference to Cicero, and even then, without a citation—Humboldt compensates for this in the paragraphs that follow by furnishing data on women's languages from around the world. This section of the text looks noticeably different, replete with extensive footnotes and references, and making use of conventional scientific language. Humboldt discusses a women's language of Japan, but he is most interested in the Weibersprachen of the Americas, even going so far as to have compiled his own dictionary of Carib women's words out of the material provided by Raymond Breton.[114] Humboldt acknowledges that the universal manner of female speech he describes above and the women's language of the Antilles are not necessarily related—returning to the raptio explanation, he maintains that the existence of the Carib Weibersprache "belongs, as most have assumed until now, more to history than to the study of language."[115] Humboldt's claim for a distinction between the universal female manner of speech and the Caribbean women's language is contradicted, however, by the connection that his own text repeatedly generates. The very structure of his text belies this distinction: by situating the two phenomena alongside each other, Humboldt signals their connection. Indeed, Humboldt's discussion of the general qualities of female speech, cited above, is the *introduction* to his section on ethnographically documented women's languages.

When Humboldt discusses other Native American languages with female variants, moreover, he does in fact connect them to a universal difference of female speech. Sex differences in these languages, Humboldt writes,

> refer, for the most part, chiefly to the naming of various degrees of kinship; but these are almost entirely different according to the sex of the speaker, which presumably derives from the difference in sentiment with which both sexes include the family circle. [. . .] Apart from this case, the *female characteristic* has no influence on the women's special manner of speech discussed here.[116]

Humboldt refers here to a linguistic prohibition in which women do not speak the names of certain male family members. Unlike Lafitau, who explained this kind of phenomenon as the result of an act of revenge, Humboldt suggests that it derives from the "female characteristic" that governs all women's speech. Humboldt does not explain how this works, but perhaps he does not need to; the work has already met his established criteria. In the internal logic of his text, facts about "primitive" Weibersprachen function as empirical evidence for his abstract and universal claims about women's speech.

That Humboldt, in documenting the vocabulary of various Weibersprachen, has gained access to something from which men would ordinarily be excluded lends this part of the text a certain voyeuristic quality. There is a section shortly before the paragraph cited above, in which Humboldt, lamenting a lack of information, describes the historical women's languages that he believes surely must have existed: "It is therefore astonishing that, as far as I know, this has not been suggested anywhere about the gynaecees of the Greeks and the harems of the Orientals. *But this may only be due to a lack of observation.*"[117] As Humboldt recounts it here, women develop separate linguistic forms in order to keep what they say secret from men. Humboldt wishes he could penetrate this secrecy and observe, for instance, the women of a harem. He remains coy about what exactly women would hope to keep "unintelligible," yet he almost certainly means words relating to sex. The notion that women establish a separate language to discuss (heterosexual) sex comes to be the most popular theory of Weibersprache in the later nineteenth and early twentieth centuries. For what else could women have to talk about? Caught up in phallocentrism, the male imagination can only imagine women talking about men, praising—or deriding—their sexual potency.

In its implicit reference to sexuality, as well as the way it brings together linguistic data with broad, speculative claims about female speech, Humboldt's text marks a radical departure from the earlier ethnographic discourse on the Carib women's language, which understood it to be yet another example of the savagery of the Americas. Beginning around 1800, Weibersprachen are increasingly understood in terms of sexual rather than cultural difference. Or to put it differently, sexual difference comes to be figured *as* cultural difference: women's distinct language is a consequence of women's distinct culture.

The Euphemistic Language of Women

Women's languages of non-Western cultures are also marshaled in support of assertions about all women's speech in *Das Weib in der Natur- und Völkerkunde* (translated into English as *Woman: An Historical Gynaecological and Anthropological Compendium*).[118] *Das Weib*, an expansive, two-volume study by the German anthropologist and gynecologist Hermann Heinrich Ploss, documents women's sexual development and corresponding rituals around the world. Chapters cover topics such as "Woman's external sex organs from an ethnographic perspective," "Menstruation in folk belief," and "The psychological constitution of woman." Ploss first published *Das Weib* serially, then as a complete book in 1885; the text was later revised by three different men between the 1890s and the 1920s.[119]

The premise of Wilhelm von Humboldt's interpretation of Weibersprache— that there is a universal female manner of speech, the product of a distinct female "characteristic"—appears in Ploss's *Das Weib*, although in less abstract terms. Ploss begins his chapter on *Frauensprache*[120] with the following assertion:

> There is a very strange phenomenon in the lives of some peoples that we cannot pass over here in silence [*mit Stillschweigen übergehen*]. It consists in the fact that among them women use their own language, which men never use and sometimes don't even understand. However, we are able to recognize different gradations of this quite clearly.[121]

There are two key points of interest in this paragraph. The first is Ploss's assertion that he *will not keep silent* on the topic of women's language; I will return to this shortly. The second is the claim that Frauensprachen are to be found in gradations. In other words, Frauensprache is a widespread phenomenon that is present to different degrees in different cultures. The paragraphs that follow, in which a panoply of women's languages is introduced and cataloged, work to confirm this assertion. From Malaysia to the Caribbean, from eastern Africa to Brazil to northwestern Canada—Frauensprachen appear to have been documented in all corners of the world. Various explanations are provided, including the familiar story of mass murder and mass rape. But Ploss does not favor this interpretation for the majority of cases. Moving quickly past the abduction justification of the Carib women's language, he argues that most women's languages are rooted in a feeling of shame about sexual matters. This applies to the language of European women, too:

> Something that could be counted as women's language can even be demonstrated among us. We only need to point out that our women also have their own manner of expression for everything connected to the sphere of sexual life, which is quite significantly different from that of men and, not infrequently, cannot even be understood by the latter. Zache reports something similar among Swahili women; the vulva is called, for example, courtyard [Hof], clam [Muschel], woman [Weib]. Here it was undoubtedly the feeling of shame that prescribed and invented these particular expressions. But we will also have to add the ban on pronouncing the names of male relatives to what is produced by the feeling of shame.[122]

Is it a coincidence that Frauensprachen are said to develop out of the "sphere of sexual life," the same topic of Ploss's book? According to Das Weib, women use replacement terms for words related to sex because they feel shame and a need for secrecy. Ploss makes it clear that he does not operate under the restrictions of such linguistic taboos: "I observe woman in her mental and physical being with the eye of an anthropologist and physician," he writes in the preface.[123] Ploss, who has no weibliche Scham (vulva)—and thus no Schamgefühl (sense of shame)—can name what women cannot. He can say genitals and sexual organs where, according to his text, Swahili women must say clam, and where German women apparently must take recourse to one of their "own expressions." And say it he does: the text is replete with descriptions—and later, diagrams—of the female body, particularly the female genitals. Thus when Ploss writes, at the beginning of the section on Frauensprache, that he "cannot pass over [this] in silence," the text inscribes a fundamental difference that between itself and its object of study. In contrasting its precise language with the metaphorical language of women, Ploss's text performs its own authority and the authority of its discipline. It participates in what Miyako Inoue, in her groundbreaking study of the construction of "Japanese women's language," calls a "network of cultural practices of objectifying femaleness/femininity and mapping a reified gender binary onto the sounds, figures, manners, and organizations of talk."[124] The notion that the physician's own language is fundamentally nonmetaphorical is, of course, one of *Das Weib*'s many fantasies. "We believe that we know something about the things themselves when we speak of trees, colors, snow, and flowers," Friedrich Nietzsche chides in an essay published around the same time as *Das Weib*, "and yet we possess nothing but metaphors for things—metaphors which correspond in no way to the original entities."[125] This does not concern Ploss's text, however, which offloads the problem of metaphoricity by attributing it to women's speech, exclusively.

In the late nineteenth century, the topic of women's rights was a pressing issue in Germany. The so-called *Frauenfrage*, or "woman question"—whether women are fundamentally different than men, and if so, whether this difference should preclude their equality—was a topic of serious debate in a variety of arenas, including medical, educational, and political discourses. To grant women legal rights and access to higher education was often taken as a menace to social institutions. Many feminists, for their part, often also insisted on woman's distinct nature, yet argued that this made women uniquely qualified for certain rights and social roles.[126] The iconoclastic thinker Lou Andreas-Salomé, for example, argues for a fundamental distinction between men and women in her 1899 essay "Der Mensch als Weib" (The human being as woman). Here, Andreas-Salomé contends that the two sexes are best understood as "two ways of life" and "two independent worlds."[127] Whereas femininity is more "primitive," harmonious, restful, and self-contained, masculinity is marked by a constant forward striving, specialization, and theoretical clarity. Women may be less inclined toward logic and "following an ever-straight line," she writes, but this makes them more in tune with reality. For "the essence of things is ultimately not simple and logical but multiple and alogical,—woman has a special resonance with this truth and involuntarily thinks individually, from case to case, even if she is logically trained."[128]

Although Ploss maintains that his book cannot be an *answer* to the Frauenfrage, he engages explicitly with the issue of women's rights at several points. *Das Weib* objects to granting equal rights to women and quotes from a number of sources to support this argument.[129] The text then concludes: "The mistakes that are made in the modern upbringing of women not only threaten their physical and moral prosperity, but are also associated with serious disadvantages for the well-being of the family and thus for the well-being of society."[130] Ploss is explicitly opposed to educating women in the same way as men, as well as to allowing women to assume occupations that would distract them from their roles as wives and mothers. One way to argue against women's emancipation is to argue for women's alterity—"Woman is in no way equal to man, but instead completely different."[131] This tactic is put to use in *Das Weib*: throughout the Frauenfrage section, Ploss's editors intersperse pictures of "exotic-looking" women between textual passages concerning the topic of women's rights. These pictures—for instance of a "Brahmin girl from Bombay, with rings in her earlobe and the edge of her outer ear" and a "Turkish woman from Constantinople in her harem," as the captions explain—have no direct relation to the text that surrounds them. Instead, they function as a kind of evidentiary substitute. The cultural difference embodied by these figures is linked to their sex and made to illustrate the alterity of European women, just as, in the section on Frauensprache, an abundance of data on "primitive" women's languages functions as evidence for claims about the difference of European women's speech.

The supposed link between women and "savages" was not the invention of *Das Weib*, but a popular assertion of the era. In *The Descent of Man* from 1871, Charles Darwin had argued that man attains a "higher eminence, in whatever he takes up, than woman can attain—whether requiring deep thought, reason, or imagination."[132] Many of women's "mental faculties," Darwin contends, are "characteristic of the lower races, and therefore of a past and lower state of civilization."[133] The perceived structural similarity between women and "lower races"—in other words, the analogy between race and sex/gender—was so pervasive in the nineteenth century that it was, in the words of Nancy Leys Stepan, "almost a cliché of the science of human difference."[134] Anatomists, anthropologists, and evolutionary biologists asserted physiological similarities between women and "primitive peoples," using "racial difference to explain gender difference, and vice versa."[135] Hence the example of the scientist Carl Vogt, who argued in 1863 that the female skull "in many respects approaches that of the infant skull, and even more so that of the lower races."[136] Here, Vogt was drawing on the work of one Hermann Welcker, who, in the previous year, had "proven" through various measurements that male and female skulls are so different that they could be considered to belong to two different species.[137] Indeed, nineteenth-century scientists often "claimed that women's development had been arrested at a lower stage of evolution," Londa Schiebinger observes.[138] As Schiebinger has also shown, however, the posited connection between racial and sexual difference was premised on the comparison of white European women to Black men; when (white, male, European) scientists considered women of color, they located them even lower in the constructed hierarchy of human development.[139]

Although Ploss notes the diverse treatment of women among the world's nations, he ultimately suggests that this variety of cultural difference pales in comparison to the difference of women themselves. Indeed, the very title of his text makes this clear, asserting *woman* as an unvariegated and monolithic category. While *Das Weib* may present itself as a book about women among the so-called *Naturvölker* (primitive peoples), it is also a book about the political status of women in Europe. To connect European women's speech to the Frauensprachen of "primitive" cultures is to undermine the contemporaneous women's movement's campaign for equality. That men and women use language differently is another argument for the naturalness of the distinction between them and the preservation of their "separate spheres."

To conclude this section, I would like to consider two more examples of how "primitive" women's languages were exploited to construct the alterity—and thus the inequality—of European women. Specifically, I am interested here in works by Georg von der Gabelentz and Fritz Mauthner. Theirs are texts that, like those of Ploss and Humboldt, mediate between empiricism and abstract essentialism in their claims about women's language use.

In 1891, Gabelentz, a German linguist known especially for his work on Chinese, issued *Die Sprachwissenschaft: Ihre Aufgaben, Methoden, und bisherigen Ergebnisse* (Language science: its purposes, methods, and previous results). As the title suggests, the book was intended as an introduction to the discipline. It also includes an analysis of women's language, which is intimately bound up with Gabelentz's own construction and defense of language science. Gabelentz devotes a good portion of *Die Sprachwissenschaft* to outlining the purview and mandates of language science, distinguishing it from other fields that take language as their focus. According to Gabelentz, linguistics (he uses the terms *Linguistik*, or linguistics, and *Sprachwissenschaft*, or language science, interchangeably) meets the requirements for a new science because its subject belongs to it alone. Language science is concerned neither with the physiology of sound production (as is natural science), nor with man's ability to structure thought (as is psychology), nor with the ability of language to express concepts and thoughts (as is logic or metaphysics), nor with the main components of the human capacity for language (as is philosophy).[140] Sprachwissenschaft is also not the same as the disciplines that study individual languages, such as Indogermanics or Hamito-Semitics,[141] nor is it solely interested in uncovering the histories of various groups—that would be *Volksgeschichte* (national/ethnic history).[142] Instead, he contends, Sprachwissenschaft is the only discipline that is interested in language as such.

While Sprachwissenschaft follows the inductive method of the natural sciences and can be compared to the natural sciences because "nothing resembles an organism more than human language,"[143] Gabelentz continues, linguistics is by no means *subservient* to the natural sciences. Like Humboldt (whom he admires but also criticizes for not being precise and empirical enough in his writing),[144] Gabelentz asserts that the truths uncovered by Sprachwissenschaft are of a higher order than those uncovered by the natural sciences. This is because knowledge about language has profound implications for knowledge about the structures of human life: "The religions, law, customs, in short, the entire cultural life of peoples is determined by the same forces as their languages."[145]

In addition to differentiating linguistics from other sciences, Gabelentz is also at pains to distinguish modern language science from the kinds of language study that came before. The modern discipline, Gabelentz asserts, takes linguistic information compiled in previous eras and renders it rigorous. Just as bear pens and menageries were the forerunners of zoological gardens, and palace displays and curiosity cabinets were the precursors to ethnological museums, linguistic information gathered by earlier ethnographers and missionaries must now "be collected for scientific research."[146] In Gabelentz's account, the contours of linguistics are clear: it is a discipline that requires "neither connoisseurship nor profound speculation, but only diligent collection and sober analysis."[147]

When Gabelentz articulates the goals of his book in the preface, he uses metaphors of geographical discovery and sexual conquest. Sprachwissenschaft, a discipline whose territories "still offer the allure and danger of virgin soil,"[148] is a project that demands the exactitude of the cartographer: "I have circumscribed [umschrieben] the whole area as far as I judged it, and I made extensive use of the cartographer's right to draw his graticule through the terra incognita."[149]

In contrast to the apparent precision of the cartographer-linguist (an unreflected image, which I will discuss more below), women's speech is, according to Gabelentz, marked by metaphor and indirectness. As he explains in his later section on Frauensprachen, women's speech is regulated by a custom that requires "that certain things [gewisse Dinge] not be called by their own names, but rather by paraphrastic [umschreibend], suggestive, borrowed names."[150] Gabelentz presents this as a general phenomenon: "Men and women each have special words that they hide from the opposite sex."[151] Although he mentions men and women here, there is no discussion of men's language in the passage that follows. Instead, it is female speech, presented as the deviation from the norm, which is Gabelentz's focus. According to Gabelentz, women have a secret vocabulary that they keep hidden from men because their Keuschheitsgefühl (feeling of chastity) "forbids one from speaking casually to members of the opposite sex about certain things [gewisse Dinge]."[152] That this linguistic division of the sexes is offered as a universal fact is not surprising. In the eighteenth and nineteenth centuries, women's chastity was considered so natural that scientists also extended it to female animals, redrawing earlier images of female apes to render them more demure and have them cover their so-called "secret parts."[153]

Ironically, in repeatedly using the phrase "gewisse Dinge" to name the sexual topics that he most certainly means, Gabelentz performs precisely that which he relegated to the opposite sex: euphemism. The *Umschreiben* of the cartographer has come dangerously close to the Umschreiben of women. Throughout his section on Frauensprachen, Gabelentz toggles between these two modes, euphemism and exactitude, buffering his vague claims with empirical data. While his generalized statements about women's speech are unsubstantiated, his claims about "primitive" women's languages are corroborated through information and scholarly references. He notes specific details about the women's languages found, for example, in the Caribbean, South America, and Greenland.[154]

As we have seen, Gabelentz engages in a laborious construction of Sprachwissenschaft as a scientific discipline. His opening chapter establishes a kind of citadel of defensive language, safeguarding the field from any attempt to dilute or disempower it. Juxtaposing Gabelentz's painstaking construction of Sprachwissenschaft as a *precise* science with his noticeably *imprecise* discussion of Weibersprachen suggests that women's language functions as both a foil and a threat to scientific discourse. Gabelentz discusses women's language not only because it is a curiosity, tied to strange peoples and their strange customs, but also because discussing it is a means of categorizing and thereby containing it—possessing the alluring and dangerous "virgin soil." Gabelentz's conflation of women's speech, in general, with women's languages, in particular, functions to reconcile his ambiguous claims with the scientific demands that he has established for his own text. And what's wrong with a little conflation? After all, Gabelentz's text implies, the experience of being woman is common across nations and throughout history.

Like Gabelentz, Fritz Mauthner devotes several pages to the topic of women's language in volume 1 of his *Beiträge zu einer Kritik der Sprache* (Contributions to a critique of language) from 1901. Mauthner's project is not philosophy or linguistics per se, but rather a critique of these disciplines and their confidence in language's ability to convey meaning. As Linda Ben-Zvi paraphrases, Mauthner's language skepticism asserts, among other things, that thinking and speaking are one activity, language and memory are synonymous, all language is metaphor, the ego does not exist apart from language, and communication between humans is impossible.[155] This understanding of language renders Mauthner's project paradoxical—How can you write an effective critique of language, when the only medium at your disposal is the same ineffective language that you aim to critique? Mauthner is keenly aware of this problem. As he writes in the introduction, the book is the culmination of a yearslong self-deception in which he attempted to "write a book against language in a rigid language."[156] Failure is immanent:

> So the decision had to be made to either publish these fragments as fragments or to hand the whole thing over to the most radical redeemer, the fire. Fire would have brought calm. But man, as long as he lives, is like living language and believes that he has something to say because he speaks.[157]

While Mauthner insists here on the fragmentary nature of his project, the overall form of the work—three volumes, with clear subject divisions, and references to comprehensive philosophical and scientific traditions—belies this description. Mauthner's construction of authority, his (in)ability to make assertions and be understood, is a fundamental tension of the *Beiträge*. This tension is never really resolved, only repeatedly performed, developed into what Katherine Arens calls a "philosophical attitude" rather than a philosophical system.[158] While Mauthner is often ambivalent about the veracity of his claims, there are also many sites where he stages his own expertise. This includes the relationship he maintains between himself and other thinkers, and his construction of Weibersprache.[159]

Mauthner begins his discussion of Weibersprache by considering the women's language of the Caribbean. He reviews the established raptio explanation, which holds that women pass on their language—a form of Arawak—to their daughters. Mauthner writes,

> In itself such a separation of inheritance by sex would not contradict generally known natural processes. We are so familiar with the fact that the rooster is always like the rooster, the hen like the hen, that we are no longer surprised by it. We are not surprised that the rooster "sings" differently than the hen. Nor that only male songbirds practice the art of singing, that only they inherited the art. The similarity of the sexes and the inheritance of characteristics according to sex are so general that it should perhaps be surprising to say that one of the greatest natural wonders is hidden in it. In any case, the separation of language according to the sexes, as it has been observed among Caribbean women, forms an analogy to one of the most common natural processes.[160]

According to Mauthner's logic, it is understandable that Island Carib mothers pass their language to their daughters because this is what happens with all species. Sex-specific characteristics, including speech, are an inheritance from one's ancestors—sexed beings beget sexed beings.

Mauthner's use of animal examples to discuss the human voice is not particularly new. In the late eighteenth century, the Comte de Buffon's *Natural History* raises the fact that male animals have a lower "voice" than females.[161] Over a hundred years later, in 1878, the French biologist Gaëtan Delaunay connects Buffon's observations of male and female animals to studies of the human larynx, ultimately reaching the conclusion that "in all species, the voice of the female is higher and weaker than that of the male [. . .] Likewise woman has a higher voice than man" and "in the lower human races, woman also has a higher voice than man."[162] Darwin also discusses differences in vocal cord length between the sexes in humans and apes, tying this difference to the "long-continued use of the vocal organs by the male under the excitement of love, rage, and jealousy."[163] Yet while Darwin and Delaunay ground their discussions of women's voices in the difference of their speech organs, Mauthner argues for a kind of difference that, while also "natural," cannot be so easily measured.

Where, precisely, does the difference of women's language reside? What distinguishes women's language from men's language—which is, as Mauthner writes in volume 3, basically the same as language itself, "no different from how our law is men's law"?[164] Although Mauthner repeatedly insists on a significant distinction between men's and women's language, he has difficulty explaining this distinction in precise terms. Women's speech diverges from men's not only in pitch and vocabulary, he asserts, but also in a general—and elusively unquantifiable—way, rendering the ultimate difference of women's speech "in many ways so difficult to express conceptually." It is a difference that the scholar can *feel* rather than observe and document: "Even among us, especially in the cultivated circles of civilized countries, there is a *palpable* difference between men's language [*Männersprache*] and women's language [*Weibersprache*]."[165] Later in the same section, he makes the declaration, "Of course, the Greeks and Romans did not pay attention to the difference between men's and women's language in their plays. [. . .] Only when women's parts were regularly played by women did it become part of the technique of drama to let women speak their women's language [*die Weiber ihre Weibersprache reden zu lassen*]"[166] and "It has been said a hundred times that you can recognize the women's language in a novel by a female writer."[167] These cryptic statements are neither explained nor elaborated.

As was the case with Humboldt, Ploss, and Gabelentz, such an ambiguous definition poses a problem for Mauthner's text, which states at the outset that, as a critique that takes language as both its subject and its medium, it must "examine the concepts much more precisely than is done elsewhere."[168] What to do, then, with a topic that is "so difficult to express conceptually" as the difference of women's speech? Mauthner's *Kritik der Sprache* addresses this problem by discharging the ambiguity of the term *women's language* to women's language itself.

Even as Mauthner attempts to describe Weibersprache in more detail, the term remains unclear. Sex-based language differences, he writes, have something to do with men's and women's dissimilar social roles and levels of education, but this cannot be the sole explanation. Among working-class people, for example, where schooling and life experience are relatively similar between the sexes, men and women still speak differently. Thus Mauthner must find another way of justifying the fundamental linguistic difference between the sexes that he asserts. He turns to a new rationale, the "chastity of women's ears," which mandates that women avoid strong or vulgar language: "Custom allows women to hear the words and even smile at them; but it forbids them from using them."[169] Yet even this potential explanation has a hazy foundation. Women's chastity may be based in nature, Mauthner writes, but may just as well be an affectation.

The only detail of women's speech that Mauthner presents with any certitude is specific to the *Salonweiber*—upper-class women who host and frequent salons. Their Weibersprache is marked by pretentiousness, which is undercut by the fact that these women do not really understand what they are saying:

> According to my observations, for example in Germany, the unnecessary use of foreign words is characteristic of this women's language. The educated male language avoids it; common people do not know of it. The woman of the salon [*Salonweib*] is just as backwards in her use of superfluous words as in her use of ambiguous words. The half-educated woman does not yet know that a certain use of French idioms can be a sign of ignorance.[170]

By characterizing Weibersprache in this way, Mauthner relieves his own text from the burden of ambiguity, displacing it onto this group of women and defining his own discourse in opposition. It is not Mauthner's critique but *women* who speak unclearly, using terms whose connotations they do not understand, and refusing to state things as they are out of prudishness. Mauthner, too, quotes in foreign languages, including French and Ancient Greek, not to mention the Latin passage from Descartes with which he prefaces the first volume, or the Italian quotation from Dante's *Paradiso* with which he concludes the third. Evidently, the difference between Mauthner and the Salonweiber is that Mauthner includes these foreign quotations necessarily and with full understanding of their meaning. Moreover, in contrast to the *Keuschheit der Frauenohren*, Mauthner does not shy away from vulgar discourse:

explicit discussions of topics such as the origins of the words *kacken* (crap) and *pissen* (piss),[171] popular names for prostitutes,[172] and the sex drives of animals and humans punctuate his text.[173]

As they appear in Mauthner's work, women have a naive faith in their ability to say what they mean and mean what they say. This is precisely the attitude that Mauthner opposes through his critique, which he describes as a practice of "unmasking," "proving to the owner of fine phrases that he possesses nothing."[174] While Mauthner's stated purpose in discussing women's languages is to provide another example of the fact that "no two groups of people speak the same language,"[175] it also serves another function—to assert the authority of his own text, even as his language skepticism calls this authority into question.

"Women's Language" into the Twentieth Century

In the early twentieth century, the universalizing impulse exhibited in Mauthner's text grows stronger, as the difference of female speech continues to be expanded to Western as well as non-Western women. This also means that the "abduction" theory loses favor, and texts on women's languages characterize this explanation as naive and outdated.[176] A 1912 review of recent German scholarship on women's languages, for example, contends that "Some other social *motif* than wife-stealing, or the capture of women, must be looked for at the source of 'women's languages.'"[177] Also around this time, several German-speaking anthropologists call the relevance of this narrative into question.[178] Whereas in nineteenth-century texts the raptio explanation was still entertained, although downplayed, in the twentieth century it is dismissed altogether. New theories are posited; the phenomenon of women's language is made to underwrite new ideologies. The Danish linguist Otto Jespersen, for example, suggests an evolutionary origin to women's languages, arguing that men and women speak differently because of the historical division of labor.[179] According to Jespersen, the work of prehistoric man, such as hunting, required silence, while woman's more domestic tasks allowed for idle chitchat—thus the development of women's "volubility." An essay in the journal *Imago*, on the other hand, claims that women's languages developed out of "repressed sexual tendencies," including the incest taboo, as theorized by Freud.[180] Although these two interpretations point to Weibersprachen in order to substantiate different theoretical programs, what they have in common is that their understanding of women's language presupposes an absolute difference between the sexes, a difference bound neither by culture nor history.

In these various cases, the act of designating Weibersprache as an object
of analysis functions to domesticate its presumed alterity. Whatever threat
women's language is imagined to pose—in its secrecy, its unruliness, its excessive metaphoricity *or* literalness, depending on who is defining it—this
threat is defanged through the imposition of scientific order. Collected and
categorized in the annals of the human sciences, the spectacle of Weibersprache becomes a mere curiosity. Yet if we look closely, we see the cracks in
the theoretical edifice. Most acutely, there is the tendency of descriptions of
Weibersprache to resemble Weibersprache itself—as in Gabelentz's inability
to exorcise euphemism from his own writing, or in Humboldt's fundamentally imprecise definition of wherein the difference of Weibersprache resides.
The texts purify their discourse by ascribing these problematic characteristics
to women, who make a useful foil. But such a move is only possible under
the star of sexual complementarity, only licensed by the fantasy of woman's
natural and necessary subordinance.

The Sex of Language:
Grammatical Gender

The previous two chapters pinpointed a discursive shift around 1800, where language became aligned with masculinity in German linguistic thought. In philosophies of language origin, the posited first language was now exclusively male, illustrated through male characters and defined through masculine properties. Something similar occurred in the interpretation of "women's language." Whereas "women's language" was earlier understood as an index of savagery, in the nineteenth century it was remade into an inevitable symptom of women's alterity, a deficient version of regular (men's) language. In both of these cases, the assumed validity of the sexual hierarchy, as a sine qua non for the conceptualization of language, functioned to turn fantastical speculation into philosophical and scientific truth.

This chapter examines a different site where the conceptualization of language and sex converged in this period: the theory and analysis of grammatical gender. For many nineteenth-century scholars, grammatical gender was a prized form. As we saw in the introduction, the linguist Wilhelm Bleek links grammatical gender to scientific progress, arguing that only people who speak "sexual languages" (his term for languages with grammatical gender) are capable of intellectual innovation. Here, Bleek builds on the idea, popular in linguistic theories of his time, that languages with grammatical gender are closer to a "pure" form of language than those that use other means of nominal classification. Several decades earlier, Wilhelm von Humboldt had esteemed grammatical gender as a structure through which language animates the world. According to Humboldt, nouns should have genders because language is an organism, and organisms are sexed.[1] Just as language consists of both "masculine" active form and "feminine" passive material but is most authentically identified with masculine activity, so too does the masculine grammatical gender become language's preeminent category in nineteenth-century accounts. This is especially evident in the work of Jacob Grimm. For Grimm, grammatical gender is an eminently sensible structure; one must simply suss out the relationship between a noun's referent and various sexed characteristics. Grimm writes, for instance, that *hand* is feminine while *fuß* (foot) is masculine, because hands are smaller and more delicate than feet.[2] In the Germanic languages, moreover, *stamm* (tree trunk) is masculine because a trunk is the "father and sustainer of the entire tree" (*DG* 410), while many words for insects are feminine because insects are small and weak (*DG* 365). Grimm studies grammatical gender in order to understand the historical development of Germanic languages like Gothic and Old High German. Yet he also, along the way, affirms gender as a necessary expression of sexual difference in language. Gender is an "application or transference, already at work in language's earliest state, of natural sex to each and every noun" (*DG* 317).

By insisting not only on an original connection between grammatical gender and biological sex, but also that this connection is fundamentally *intelligible* to the modern speaker, Grimm exemplifies an approach to grammatical gender that emerges in the late eighteenth century. In this way, the interpretation of grammatical gender undergoes a shift similar to those outlined in chapters 1 and 2. Whereas Enlightenment grammarians consider gender to be a capricious structure with little semantic value, scholars from the 1770s onward argue that grammatical gender has a legible meaning, easily deciphered through the stable categories *masculine* and *feminine*. As nineteenth-century linguists attempt to make sense of grammatical gender, they employ many of the tropes that undergirded theories of language origin and explanations of women's language, such as the primacy of man over woman and the absolute necessity of sexual differentiation for any conception of human life. What the discourse on grammatical gender shows in particular—and what I will concentrate on here—is how linguistic thought from this period not only *presupposes* ideas about the differences between men and women, but also turns to language to *shore up* these differences. In discussions of grammatical gender, the relationship between language and sex appears as a tautology: the hierarchy of the sexes is both a cipher, the universally valid key through which linguistic data is interpreted, and what that data is made to corroborate.

Grammatical Gender as Capricious Category

Before turning to the theories of Humboldt and Grimm, I would like to provide a general definition of grammatical gender, as well as an overview of how it was understood directly prior to the formulation of their theories. I restrict my discussion to the seventeenth and eighteenth centuries, since an exhaustive account of the history of thinking about grammatical gender is outside the scope of this book. Interested readers may consider Anthony Corbeill's excellent study *Sexing the World* to learn about theories of grammatical gender in ancient Rome.[3]

In the most basic sense, grammatical genders are "classes of nouns reflected in the behavior of associated words," to use the classic definition from Charles Hockett.[4] To qualify as a gender system, "every noun must belong to one of the classes, and very few can belong to more than one."[5] Languages can have various numbers of classes. French, for instance, has two genders (masculine and feminine), while German has three (masculine, feminine, neuter). Although English does not have a strict gender system like German or French, it does encode gender through pronouns.[6]

In the theories of grammatical gender examined in this chapter, the origin of *animate* noun genders rarely comes under scrutiny. Scholars understood there to be an intuitive connection between sexed beings and the genders of the words that represent them. It was generally not questioned, in other words, that nouns for *woman* (such as *femme* or *Frau*) should be of the feminine gender, while nouns for *man* (such as *homme* or *Mann*) were masculine. This assumption also extended to some animals, such as *Katze* versus *Kater* (cat vs. tomcat). More difficult, however, was explaining why the nouns for inanimate objects, which have no sex, were not always neuter, but could also be masculine or feminine.

Grammarians of the seventeenth and early eighteenth centuries addressed this problem by asserting a general connection between grammatical gender and biological sex. A language's first speakers—so the argument went—either personified inanimate objects as living, sexed beings or associated them with qualities found in the male and female sexes. A noun's gender was the transfer into language of the sex assigned to or associated with its referent. Different explanations were proffered for the neuter—it designated an object either understood to have no sex, whose sex was not known, or whose sex was neither male nor female. Regardless of the specific differences in their theories, most scholars from this period argued that whatever had motivated early man to classify one object as masculine, another feminine, had to remain obscure, since this classification did not follow any recognizable logic. For the modern speaker, these texts suggest, noun genders have little meaning or value.

The influential *Grammaire générale et raisonnée de Port-Royal* (Port-Royal General and Rational Grammar) from 1660, for example, maintains that, except for nouns that refer specifically to men and women, or to objects associated with them, there is no relation between a word's meaning and its gender.[7]

Grammatical gender was created "by pure caprice and a habit without reason. This is why such determination varies according to different languages, and even in the words which one language has borrowed from another."[8] Around the same time, J. G. Schottelius, in his *Teutsche Sprachkunst*, does not even entertain the idea that grammatical gender could have semantic value. Although Schottelius provides his reader with categories of nouns that are typically masculine or feminine, such as "the names of women/ female occupations/ female vices / trees/ tree fruits/ rivers," he includes no explanation of why rivers and tree fruits, for instance, are of the feminine gender in German.[9]

This assumption—that the motivation behind noun genders is inscrutable and better not dwelled upon—recurs in the first half of the following century. In 1747, the Abbé Gabriel Girard argues that there is indeed a connection between grammatical gender and biological sex: masculine nouns refer to objects understood as having "a relation [*raport*] to the male sex," feminine nouns likewise having "a relation to the female, or belonging to this latter sex."[10] Yet, Girard continues, this relation does not follow an identifiable system. When humans first assigned words their genders, they did so "without consulting either Logic or Physics. That which the first stroke of imagination painted without examination, Usage confirmed without deliberation."[11] According to Girard, gender is "an incidental idea," created "with neither motive nor systematic design."[12] Similarly, the 1757 entry for gender in the encyclopedia of Denis Diderot and Jean le Rond d'Alembert asserts that many inanimate nouns were assigned their gender either arbitrarily or based solely on formal characteristics: "Some words will be of a certain grammatical gender because of the sex, others because of their ending, a great number of the basis of pure whim."[13] Thus it would be "an unnecessary trouble, in any given language, to seek or to establish rules to make words' genders known: only usage can provide such knowledge."[14]

Other Enlightenment-era texts engage even less with the topic of gender's origins, favoring the more practical concern of standardizing grammatical gender, which could vary by dialect, over speculating about its genesis. In his *Grundlegung einer deutschen Sprachkunst* (Foundation of a German language art) from 1748, for instance, Johann Christoph Gottsched does not entertain a theory of grammatical gender's origins except to say that there is some correspondence between grammatical gender and biological sex.[15] Of course, Gottsched explains, there are the words that correspond to sexed beings: one should use the feminine article for women's names, for words referring to women, and for female occupations.[16] As for all other nouns, however, grammatical gender does not necessarily relate to meaning: it must either be derived from the form of the word (e.g., all words ending in in the suffix -*heit* are feminine) or simply memorized.

In the late 1700s, however, something changes in the way grammatical gender is discussed. Scholars begin to assign the form a semantic value as they speculate about the motivation behind the genders of various nouns. In 1774, for instance, Antoine Court de Gébelin writes that a word was given the feminine gender "whenever its object presents some qualities of the female sex; when it has more grace than strength, more softness than intensity, more gentleness than vigor, or if it is a Being capable of reproduction and impregnated by Nature; and more passive than active."[17] This leads to delightful yet wholly unsubstantiated claims like the assertion that the word *mer* (sea) is feminine because it was understood to be "the container and the generator of a tremendous quantity of plants & animals," whereas *océan* (ocean) is masculine because it captures "its immense range and the dreadful roaring of its waves."[18] In the 1780s, Johann Christoph Adelung likewise makes an argument for a direct link between *genus* and *sexus* in inanimate nouns. According to Adelung, masculine nouns correspond to "everything that had a concept of liveliness, activity, strength, size, and also of the dreadful and the terrible," while feminine nouns are that which "were thought of as receptive, fertile, gentle, passive, pleasant." Neuter, on the other hand, has two possible explanations: that "in which one didn't become aware of such things, or at least did not want to express them."[19] Two decades later, August Bernhardi would provide a similar justification for grammatical gender, explaining that in German, the names of winds and rivers are masculine "because of their remarkable power," whereas the names of trees, cities, countries, and islands are feminine because they are associated with containment and receptivity.[20]

In the following three sections, I chart the changing imagination of grammatical gender between 1770 and 1890. Scholars in this period increasingly esteem grammatical gender as a necessary and beneficial form—a valorization that, as we will see, goes hand in hand with an assertion of the complementary model of the sexes as absolute truth. Not only does this move construct male dominance as natural and inevitable, but it also makes possible sweeping claims of cultural, and at times racial, superiority. The first section examines texts by Herder, Adelung, and Karl Philipp Moritz, which argue that grammatical gender results from an essential animistic enlivening of the world through language. Section 2 analyzes theories that valorize grammatical gender from Wilhelm von Humboldt, Wilhelm Bleek, and Jacob Grimm, paying special attention to Bleek's link between grammatical gender and "scientific progress" and Grimm's assertion of the primacy of the masculine. The third section reconstructs a debate over the origin of grammatical gender that occurred in the 1890s between Grimm's followers, on the one side, and Karl Brugmann and the Neogrammarians, on the other. Their argument centers on whether noun genders spring from personification, and, correspondingly, whether we can use categories of biological sex to make sense of grammatical gender—whether human categories bear any relevance for categories of language. This relatively obscure debate, which has mostly been relegated to the archives of history, is a rich source for examining how the relationship between language and sexual difference was understood in this era. The arguments made therein can also be rather startling. Grimm's care for linguistic data is matched only by his enthusiasm for what Ernst Cassirer called the "power of aesthetic fantasy and linguistic empathy,"[21] and what we might otherwise call conjecture—a conjecture that takes woman's natural subordination as both its premise and conclusion.

Heterosexual Nouns

In his *Treatise on the Origin of Language*, Herder calls grammatical gender the "genitals of speech." Anthony Corbeill traces the practice of allowing "words themselves to participate in a biology of sexual reproduction" back to the Roman scholar Varro, who derived *genus* "from the verb *generare*, 'to beget,' since genders 'are only those things that give birth.' "[22] Dennis Baron follows this interpretation of the word *gender*—whose original meaning of " 'kind, or sort,' had nothing to do with sex"—through the medieval and into the modern period.[23] Herder is of course drawing on this tradition here, but it is important to note the specifics of his theory: for him, gendered nouns are "genitals" because for the first speakers, everything they referred to was alive, and everything alive was sexed. "Because the human being related everything to himself," Herder explains,

> because everything seemed to speak with him, and really acted for or against him, because he consequently took sides with or against it, loved or hated it, and imagined everything to be human, all these traces of humanity impressed themselves into the first names as well! They too expressed *love or hate, curse or blessing, softness or opposition*, and especially there arose from this feeling in so many languages *the articles*! Here everything became human, personified into woman or man—everywhere gods; goddesses; acting, wicked or good, beings![24]

In the animistic mindset that Herder describes, inanimate objects are not *like* men and women but *are* men and women. Oceans, zephyrs, and springs are transformed into "acting, wicked or good, beings." What is significant about Herder's explanation is the double binary that it asserts. Inanimate objects, once enlivened, must become not only benevolent or threatening, but also male or female. There is no personification without sexual difference; "the human" is always already sexed.

The grammarian Johann Christoph Adelung makes a similar argument for a direct link between *genus* and *sexus* in the 1780s. Adelung published three texts dealing with grammatical gender; I will focus here on the 1783 "Von dem Geschlechte der Substantive" (On the genders of nouns), which is his longest and most developed.[25] Adelung draws heavily on Herder, although adds more detail about the original act of producing grammatical gender. "Now it is also not difficult to understand," Adelung writes,

> according to what reasons these sexes were divided. Anything that resembled masculine qualities, if it involved strength, liveliness, efficacy, the sublime, was endowed with an imparting, creative and active force, was a masculine being, and its name became a masculine noun. On the other hand, if an object reveals feminine qualities, if it has more charm than strength, more gentleness than liveliness, more delicateness than strength, and generally behaves more passively than actively, then it was regarded as a female being and, consequently, its name was of the feminine gender.[26]

For Adelung, gender is a "remainder of the first childhood of the human race,"[27] an artifact of an era that the Germans have long since left behind. People no longer interact with the objects of the world by considering them to be alive; their language, Adelung asserts, should reflect this reasoned outlook. He describes the elimination of grammatical gender as a marker of enlightenment and cites English as a model language in this regard.[28] The primitive creations of grammatical gender, he suggests, were "unstable and arbitrary."[29]

It may seem difficult to reconcile Adelung's insistence on the arbitrariness of grammatical gender with his earlier assertion that the motivation behind gender is "not difficult to understand." How is it possible that the allocation of gender is both arbitrary and obvious enough to allow for modern insight? We may make sense of this apparent contradiction by distinguishing between categories of gendered attributes, which Adelung asserts as universally valid, and the way that the world is interpreted through them—which is contingent, culturally dependent. Thus Adelung explains why the same objects can have different genders in different languages:

> The sun is an active being for the Greeks and Romans, fertilizing everything through its light and warmth, for the Germans a gentle, charming, revitalizing being. For the first two the moon is feminine because its light is weaker and

gentler than the light of the sun; for the Germans, in contrast, it is masculine
because they saw it as the active cause of the most distinguished time spans,
and perhaps also because they believed in its influences on the earth and human destinies from a very early age. And likewise with other words.[30]

The inanimate objects of the natural world have no sex; they can be assigned
any gender, depending on how a culture views them. What cannot proceed
arbitrarily, however, is the classification of masculine and feminine characteristics. The truth that grammatical gender evinces is therefore not, for example, the inherent femininity *of the moon*, but rather the inherent femininity
of the *qualities* weakness and delicateness. The gendered categories employed
by Adelung are presented as unalterable and universal, uniting languages
across space and time. His theory allows many things to be historicized and
relativized, but not the conceptualization of sexual difference.

What about the neuter? Adelung distinguishes between what he calls *das
persönliche Geschlecht* (either feminine or masculine) and *das sächliche Geschlecht* (neuter). When language was first created, Adelung writes, words
could only have been classified as masculine or feminine, "since nature actually only knows two genders." The neuter gender, on the other hand, dates
from a more enlightened, historically later period, when man had dispensed
with anthropomorphizing and acquired the reason necessary to classify inanimate objects as sexless. This is a shift from what Adelung first writes on
gender in 1781, namely that the neuter was assigned to everything in which
man either did not discern a sex, or did not want to express it. Instead of
being a third sex, in which neither "male" nor "female" characteristics appear dominant, the neuter is now defined as the complete absence of sex and
linked to inanimacy.[31] The categories of human sex, and the categories of
grammatical gender created in their image, have been strictly limited to male
and female. Indeed, as he put it already in 1782, "God created nothing, it says
in the Qur'an, that is not male or female; this is certain of all the creatures of
the earth, of souls, and even of things of which one should least expect it."[32]
The possibility of a third sex, briefly entertained in 1781, is quickly discarded
in favor of a rigid gender binary.

In the years that follow, Adelung's ideas on grammatical gender are taken
up and reinterpreted by other German scholars. Bernhardi elaborates on the
theory that the neuter gender dates from a later period, strongly linking the
appearance of the neuter to the rise of reason.[33] Karl Philipp Moritz, on
the other hand, further reinforces the masculine/feminine binary as he makes
the remarkable argument that German nouns can be sorted into pairs that

mirror human heterosexual coupling. In *Deutsche Sprachlehre für die Damen* (German language lessons for women), Moritz writes that every masculine noun was created such that it has a feminine counterpart. "Hence that is now why, for example, we say,"

    der Baum, die Blume
    der Wald, die Wiese,
    der Zorn, die Sanftmuth,
    der Haß, die Liebe.

Wo denn auch der härtere, männlichere Artikel der in das sanftere die hinü-berschmilzt. So scheint die Sprache auch alles leblose in der Welt zu paaren, indem sie zu etwas Größern oder Stärken immer etwas Aehnliches aufzufin-den weiß, das nur kleiner oder schwächer, aber schöner und angenehmer ist.[34]

    The (masc.) tree, the (fem.) flower
    The (masc.) forest, the (fem.) meadow
    The (masc.) anger, the (fem.) meekness
    The (masc.) hate, the (fem.) love

Where the harder, more masculine article (der) melts into the softer one (die). That's how language seems to pair everything inanimate in the world, in that it always knows how to find something complementary to something bigger or stronger, which is not only smaller or weaker, but more beautiful and more pleasant.

Moritz, like Adelung, subscribes to a theory of animation: he writes that grammatical gender originates in the extension of *Persönlichkeit* to inanimate objects.[35] Like Herder, Moritz also grants language itself a productive func-tion: it is *language* that sorts nouns into couples, that orders the objects of the world into gendered pairs. In this way, language mirrors what is, for Moritz, a central fact of human life: the coming together of men and women in het-erosexual marriage.

While Moritz's noun couples may seem like poetic fancy, they are an ex-treme example of the linguistic heteronormativity that many activists still struggle against today. I will discuss contemporary reform efforts (and the backlash against them) in more detail in the book's Coda, but cite here one ex-ample: the work of linguist Lann Hornscheidt. In an effort to support gender-neutral, trans-affirming, and antidiscriminatory language, Hornscheidt argues against linguistic "genderism." One form that genderism takes is *Zweigen-derung* (two-gendering)—the "social normalization and standardization of two and exactly two genders as a framework of social comprehensibility or intelligibility, as a fundamental constitution of human existence."[36] For Horn-scheidt, Zweigenderung is not necessarily a function of grammatical gender but rather a consequence of how certain linguistic practices interpellate and exclude—how the address *Ladies and Gentlemen* promotes a clear binary, for example. For Moritz and his contemporaries, in contrast, language's expression of the sexual binary is both inevitable and natural.

Texts like Moritz's are distinguished not only from twenty-first-century gendered language reforms but also from earlier accounts. In the earlier 1700s, there was little interest in grammatical gender, its nature, or its origins. Neither Beauzée nor Gottsched spends much time speculating about what motivated the first speakers to assign various words their genders. But for those who come after these Enlightenment thinkers, the origin of grammatical gender is a subject of fascination: the topic "deserves a little investigation," as Adelung puts it in his introduction. There are indeed scholars who controvert this approach and criticize the idea of linking grammatical gender to biological sex. Johann Werner Meiner and Christian Heinrich Wolke, for instance, argue in 1781 and 1812, respectively, against connecting a word's gender to its meaning.[37] Yet their voices are in the minority. From the late eighteenth through the late nineteenth century, the dominant theory of grammatical gender asserts a semantically intelligible motivation, arguing that the meaning of noun genders can be accessed through the sexed characteristics of contemporary men and women.

To make sense of this newfound interest in grammatical gender, in the meaning attributed to the form beginning in the 1770s, we must recall the reconceptualization of language that takes place around 1800, which I discussed in chapter 1. Instead of being a tool to represent prelinguistic thought, developing out of instinctive natural cries, language is now understood to be what makes thought itself possible—indeed, what makes possible the human *as* human. If language is an organism, as we saw Wilhelm von Humboldt argue, and if there is a reciprocally formative relationship between the structures of this organism and the structures of human life, and, moreover, if sexual difference is a fundamental feature of human life, then language, too, must incorporate sexual difference within itself. Furthermore, in a period increasingly invested in the relationship between language and national identity, the divergent national "choices" behind the gendering of nouns in various languages were assumed to provide insight into the culturally specific mindset of its creators. As we will see in the following section, the fact that nouns have different genders in different languages was used to reveal the uniqueness—and often the superiority—of a culture's worldview.

Gender's Necessity

Whereas earlier writers such as Herder and Adelung focused on the way that "primitive man" had enlivened the world through language, transforming inanimate objects into living beings, nineteenth-century discussions of grammatical gender de-emphasize the animation of objects. They instead claim that gender exists in language because sex is as important a category for language as it is for human life. This perspective is especially evident in the work of Wilhelm von Humboldt, who investigates grammatical gender in two different texts: an 1827 letter to the Sinologist Jean-Pierre Abel-Rémusat and the 1828 essay "Ueber den Dualis" (On the dual). In both instances, Humboldt affords grammatical gender a positive value, considering it an indication of the productive potential of language. As he writes in the letter to Abel-Rémusat, gender belongs entirely to

> *the imaginative aspect of languages.* The examination of the mind and its intellectual relations could not prove it; considered from this point of view, it could easily be ranked among the imperfections of languages, as not very philosophical, as unnecessary and as out of place. But as soon as a nation's young and active imagination invigorates all the words, entirely assimilates language to the real world, and completes its prosopopoeia, by making each period a table where the arrangement of the parts and the nuances belong more to the expression of thought than to thought itself, then words need to be gendered, as living beings belong to a sex. A consequence of this is technical benefits in phrasing; but to appreciate it and feel the need for it, a nation must be struck mostly by what language adds on to thought by turning it into speech.[38]

Humboldt focuses on the word itself rather than what it represents: the imagination vivifies "all the *words*," not "all the *objects*." The linguistic structure of gender may have been inspired by the outside world, by biological sex, but it is not a simple one-to-one reflection of sex (indeed, if it were, all inanimate objects would be neuter). Instead, Humboldt understands grammatical gender to be a way of animating language itself. Words must have genders because language, too, is a kind of living being, a creative and dynamic organism.[39]

One year later, in "Ueber den Dualis," Humboldt further elaborates on this point. Here, he links sexual difference to a general principle of complementarity, a kind of ontological dualism, which he refers to as the *Begriff der Zweiheit* (concept of duality). The Begriff der Zweiheit is identifiable in all aspects of life and belongs to the realm of the visible as well as invisible. It is predominant "in the laws of thought, in the striving of sensation, and in the deepest foundations of the organism of the human race and of nature."[40] According to Humboldt, the division of the sexes is one of the first dualities that man recognizes. After he learns to differentiate between groups of two versus groups of more than two, "then men's perception and feeling of duality gives way to the division of the two sexes and into all concepts and feelings related to them."[41] This recognition of binarism is subsequently extended to all kinds of natural phenomena, including day/night, earth/sky, and land/ water. Man furthermore finds duality within the dialogic structure of language, in the fact that the very possibility of language "is conditioned by address and response."[42] The previous attempts to naturalize grammatical gender that we encountered all functioned by relating gender to sex—languages have masculine and feminine genders because humans are male and female. Humboldt takes this reasoning one step further and naturalizes the division of the sexes itself, incorporating sexual difference into a grand scheme of complementarity and duality.

Because duality is inherent to the dialogic act at the root of linguistic communication, Humboldt claims that the dual is a grammatical form well suited to the structure of language. This is a status that he also claims for grammatical gender. Not all aspects of the world are equally suited to linguistic assimilation. The distinction between animate and inanimate objects, which is used in some Native American languages to classify nouns instead of gender, does not contain anything, asserts Humboldt, "that could be merged internally into the form of language, grammatical distinctions based on them remain in language like a strange material, and testify to a dominance of the sense of language that has not been completely penetrated."[43] Humboldt implies that the distinction of sex, on the other hand, is not only a category more easily absorbed by language, but also one that brings us closer to the "pure form of language." Every language that adopts grammatical gender, Humboldt writes,

is, in my opinion, one step closer to the pure form of language than one that is content with the concept of animate and inanimate, although this is the basis of gender. But the sense of language shows its dominance only when the gender of beings is really made into a gender of words, when there is no word which, according to the manifold views of the language-forming imagination, is not assigned to one of the three sexes. Calling this unphilosophical was a misunderstanding of the true philosophical meaning of language. All languages that only designate biological sex and do not recognize any metaphorically designated gender prove that, either originally or in an era when they no longer observed this distinction of words, or when they, in confusion, threw the masculine and the neuter together, were not energetically penetrated by the pure form of language, did not understand the fine and delicate interpretation which language lends to objects of reality.[44]

Those languages that distinguish between animate and inanimate—instead of between masculine, feminine, and neuter—are, in Humboldt's formulation, further from the "pure form of language" because they do not exercise their imaginative and metaphorical capacity. To make a nominal distinction between animate and inanimate objects is simply to reflect, in language, the world as it is. To classify nouns based on gender, however, is to allow language to reinterpret and reimagine the world—rendering the sky masculine, for instance, or the sun feminine. This, writes Humboldt, is the true law of language: "That everything that is transposed into [languages] sheds its original form and assumes that of language. This is the only way that the world is successfully transformed into language, and that the symbolization of language is completed by means of its grammatical structure."[45] Humboldt values grammatical gender because it subordinates the categories of the world to the categories of language, whereas the nominal distinction inanimate/animate allegedly performs the opposite. In an essay on metaphor from the 1850s, the linguist August Pott similarly refers to the creation of grammatical gender as an "act of freedom" of the "language-creating spirit."[46]

   Humboldt does not declare a relation between sexual difference and the underlying structure of language and cognition, as he does with the dual. Yet he does, elsewhere in the essay, claim that the sexual difference, in its "most general and spiritual form," plays a fundamental role in all human thought and feeling—no small assertion.[47] Grammatical gender is thus significant for two reasons: First, because it introduces sexual difference, a central distinction of human life, into language. And second, because grammatical gender is language's *reconfiguration* of sexual difference, rather than its mere reflection.

GRAMMATICAL GENDER AND
CULTURAL SUPERIORITY

Especially in the second half of the nineteenth century, the discipline of language study was often—though not always—complicit in claims of European cultural and racial superiority. This has been documented extensively by scholars working on the history of "Aryan" mythology and on the construction of race in the human sciences. Sprachwissenschaft, Ruth Römer writes, "has been an ancillary discipline of ethnography, ethnology and anthropology for centuries, but conversely the division of languages also gave ethnologists and

anthropologists concepts that they willingly took up and adapted."[48] Tuska
Benes shows that in the mid-nineteenth century, philologists "redefined race
to accommodate a form of linguistic, rather than biological, determinism.
Racialist discourse thoroughly permeated research on the Indo-European and
Semitic language families [. . .] even if most comparativists rejected a reduc-
tion of linguistic diversity into biology."[49] To uncover the Indo-European or
"Aryan" *Ursprache* meant to uncover the "Aryan" homeland and subsequent
lineage. At stake in these investigations was the honor of belonging to a people purportedly superior to all others—the most advanced, creative, *civilized*.
While the racism of language science has received significant attention, there
has been less focus on the way that this racism intersected with claims about
the sexes. In this section, I examine how linguistics' construction of racial
categories often depended on masculinist thinking.

Bleek's 1867 *Über den Ursprung der Sprache* (*On the Origin of Language*),
which argues that grammatical gender is indicative of a higher level of civilization, offers an illustrative example. In this text, Bleek radicalizes Humboldt's
valorization of grammatical gender, asking, "Is it, then, a mere accident that
nearly all the nations which have made any progress in scientific acquirement
speak sexual languages [*sexuelle Sprachen*]?"[50] Among this esteemed sexual
class, Bleek counted the speakers of ancient Hebrew, Egyptian, Arabic, Greek,
Latin, and, naturally, his own mother tongue. In Bleek's text, the link between
grammatical gender and scientific knowledge follows a clearly presented, if
spurious, logic. As certain groups of humans extended "male" and "female"
pronouns to inanimate objects, he writes, they were led to personify those
objects as living beings, which made those humans interested in the world
around them. This personification incited "profound study," sharpening the
human "power of observation," and paving the way for nothing less than a
"knowledge of the final ground of all existence."[51]

As for nations whose languages have never had grammatical gender,
Bleek baldly claims that none has "added any noteworthy contribution to
scientific knowledge; and not a single individual who could be called great
as thinker, inventor, or poet."[52] The relationship between language and intellectual ability is mutually reinforcing. According to Bleek, nations that lacked
the wherewithal to invent grammatical gender will never develop it, as they
are stuck with grammatical forms that "do not allow their imagination that
higher flight which the form of the sexual languages irresistibly imparts to
the movement of the thought of those that speak them."[53] Predictably, Bleek's
dichotomy breaks down along racial lines: the most "unscientific" of all languages are purportedly spoken by Black Africans, the most "scientific" by white
Europeans.

Although, as Robert Bernasconi argues, the "broad division of peoples on the basis of color" has a long and violent history, it is not until the late eighteenth century—specifically in the work of Immanuel Kant—that "race" becomes the scientific concept that would inform so much of nineteenth-century thought.[54] Kant wrote about race in several essays, and in them we can see the roots of the racialized hierarchy employed by Bleek and many of his contemporaries.[55] In the 1777 "Of the Different Human Races," Kant distinguishes between four races: "white," "Negro," "Hunnish," and "Hindu-ish."[56] For Kant, the white race, "noble blond," represents the original race, while the so-called "Negro" is derivative: "Strong, fleshy, and nimble, but, under the care of his motherland, lazy, soft, and dallying."[57] Kant says nothing about "sexual languages," but the dehumanizing rhetoric that accompanies his racial hierarchy is echoed in Bleek's work. Bleek spent his career in Cape Town, where he studied and documented the languages of southern Africa. His work advanced a racist evolutionary theory: Bleek compared the peoples whose languages he researched to animals and ranked them according to their ostensible level of development.[58]

That Bleek's racialized hierarchy is implicated in his own performance of authority is plain to see. Anxious about his location "in the Southern Hemisphere, so far removed from the bustle of the European learned world,"[59] Bleek is at pains to display his scientific bona fides. In the preface, he recounts having conversed "with one of the most prominent geologists of our time" and establishes that he has not only read, but also understood, the work of Charles Lyell.[60] Bound up with the assertion of his personal authority is his defense of his discipline. Conceived as a "first chapter in a history of the development of humanity," *On the Origin of Language* is replete with references to data and calculations. Bleek even goes so far as to calculate the *date* of the origin of language—marking that epoch at which man became man—and situates it at approximately "a hundred thousand years before our usual reckoning."[61] Although he ultimately decides not to pursue such calculations, rendering unto paleontology what is paleontology's, Bleek continues to make a case, throughout the text, for the significance of linguistic analysis to the scientific study of human history. It does not hurt this case that he conducts his analyses in German, one of the preeminent "sexual languages." In this way, he defends German as a language of scholarship in the context of British colonial South Africa, where speakers of English, a language that has "altered and brought into almost complete accordance with reason, the original distinction of gender," may look with "astonishment" on the gendering of nouns.[62] As Bleek reminds his reader of the imaginative and scientific capacities that languages like German apparently facilitate, he bolsters his own identity, fashioning himself a place in the Western scientific tradition and distinguishing himself from the "unscientific" peoples upon whom his gaze has come to rest.

Bleek writes that the connection between "sexual languages" and scientific progress is *no accident*. This is a telling phrase, as his text is deeply invested in distinguishing the arbitrary from the necessary. While the contingencies of culture determine the morphology and semantics of grammatical gender in various languages, the connection between this structure and higher intelligence is a necessary universal. Bleek presents the partitioning of the world's languages into sexual and nonsexual as one of humanity's ultimate dichotomies, but it depends, in fact, upon another structure of "necessary" difference: that of man versus woman. Embedded within Bleek's descriptions of primitive forefathers and linguistically anthropomorphized objects is an argument for the centrality of sexual difference to human identity.

Around the same time of Bleek's work, the linguist Max Müller connects the existence of noun genders to the "mythopoetic" stage of the "Aryans" and their creativity "in areas such as art, politics, and science," as Stefan Arvidsson writes.[63] In 1863, Karl Richard Lepsius, a German scholar of Egypt and a contemporary of Bleek's, also makes an argument for the significance of grammatical gender. Bleek studied Lepsius's work and met the Egyptologist in Berlin in 1852; it is not clear which of them originated this particular interpretation of grammatical gender, or whether they developed it independently.[64] Nevertheless, there are striking similarities in their insistence on grammatical gender as a *necessary* indicator of cultural superiority. In the *Standard Alphabet for Reducing Unwritten Languages and Foreign Graphic Systems to a Uniform Orthography in European Letters*, Lepsius contends that it is "not accidental but very significant that, as far as I know without any essential exception, only the most highly civilised races—the leading nations in the history of mankind—distinguish throughout the genders."[65] What is it, exactly, that links grammatical gender and civilization? Lepsius says only that "the development of peculiar forms for the grammatical genders proves a comparatively higher consciousness of the two sexes." This lack of an explanation, however, does not stop him from making grand claims:

It seems however unquestionable, that the three great branches of gender-languages were not only in the past the depositaries and the organs of the historical progress of human civilisation, but that to them, and particularly to the youngest branch of them, the Japhetic, belong also the future hopes of the world. All the other languages are in decline and seem to have henceforth but a local existence.[66]

Because they are so undeveloped, he suggests, languages without grammatical gender do not merit being classified according to formal features but should instead be categorized by geographical distribution. When Lepsius revisits the subject of grammatical gender in his 1880 study *Nubische Grammatik* (Nubian grammar), he implicates the form not only in a hierarchy of the world's cultures, but also in a hierarchy of the sexes. "Since man forms language, the differentiation of genders proceeds by separating out the feminine, which is why that gender is particularly developed," he notes in the text's introduction.[67] Lepsius says here what other grammarians assume but do not bother to state explicitly: that the form of grammatical gender was created specifically by *men*, since language is a masculine creation. Lepsius's statement from 1863 that grammatical gender indexes cultural development because it evinces a "higher consciousness of the two sexes," then, refers to an awareness not simply *of* the two sexes, but specifically to the asymmetrical relationship between them. Civilization depends upon the recognition of man's preeminent, and woman's subordinate, status.

GRIMM'S PRIMACY OF THE MASCULINE

As in the work of Bleek and Lepsius, the assumption of woman's natural and fundamental subordinance to man is central to Jacob Grimm's writing on grammatical gender. If figures such as Herder, Adelung, and Humboldt establish grammatical gender as a meaningful structure, it is Grimm who speculates most extensively about what this meaning might entail. Grimm hypothesizes not only about the general origin of grammatical gender but also about the rationale behind the genders of a myriad of nouns.

Although the name Grimm is most commonly associated with the literary genre of the fairy tale—he collected, edited, and published, along with his brother, Wilhelm Grimm, the popular *Kinder- und Hausmärchen*—Jacob Grimm also made significant contributions to the study of language. In addition to his collaboration with Wilhelm on the *Deutsches Wörterbuch*, Jacob Grimm published a multivolume study of the Germanic language family, titled *Deutsche Grammatik* (1819–37). In volume 3 of this work, Grimm devotes a long chapter to the topic of grammatical gender, which he defines as "an application or transference, already at work in language's earliest state, of natural sex to each and every noun" (*DG* 317); he also describes gender as "an expansion of natural sex to each and every object, which originated in the fantasy of human language" (*DG* 346). Similar to Herder, Grimm argues that inanimate objects are gendered through language itself (the "fantasy of human language"), which takes the categories of "natural" sex as its inspiration.

Similar to prior theories, too, Grimm associates activity with the masculine, passivity with the feminine. Grimm diverges from previous scholars, however, in his addition of a *developmental* hierarchy. "The masculine," he writes, is not only the "greater, more stable, rougher, active, agile, generative" while the feminine is the "smaller, softer, quieter, passive, receptive," but the masculine is the "earlier" while the feminine is the "later" gender (*DG* 358–59). Although Grimm also discusses the neuter, his explanation of this third gender actually works to maintain a rigid binary. Grimm argues that the neuter gender correlates with "the generated, effected, material, general, undeveloped, collective" (*DG* 359)—an undeveloped or absent sex. Elsewhere in the chapter, he also suggests that the referents of neuter nouns may correspond to "a mixture or combination of active and passive, male and female forms" (*DG* 311). In Grimm's theory, in other words, there are only two "natural" genders. The neuter designates either the absence or copresence of masculine and feminine.

Grimm asserts the primacy of the masculine gender already in the chapter's introduction. The three genders did not appear simultaneously, he argues. Rather, the feminine and neuter arose from the masculine: "Although the three genders must be assumed to be present and deeply entrenched in the oldest relics of the German language and well beyond our history, the perception cannot be hereby ruled out that the masculine showed itself to be the liveliest, most powerful and most primary of all three genders" (*DG* 313). Grimm bases this declaration of the primacy of the masculine on grammatical, phonological, and semantic evidence. His claim furthermore extends to nouns designating animate as well as inanimate objects. In what follows, I will examine these different aspects of Grimm's analysis, with an eye toward the way that his understanding of grammatical gender both reinforces and depends upon the "natural" dominance of man over woman.

Let's first turn to Grimm's arguments about phonology, specifically his gendered system of sounds. Although Grimm writes elsewhere that vowels are "feminine" and consonants are "masculine" (*DG* 42),[68] he makes an additional distinction here between "masculine" short vowels and "feminine" long vowels. A weak masculine noun, writes Grimm, "is introduced by a short vowel (hana, Old German hano), that of the feminine via a long one." After a series of examples, Grimm then asserts that "it cannot be doubted that short vowels are older and nobler" (*DG* 314). Grimm does not justify this claim in *Deutsche Grammatik*, but we can explain it based on his essay on the origin of language from 1851. There, Grimm proposes the theory that language develops in several stages of increasing phonological and grammatical complexity. According to this hypothesis, in the first stage of development, language contained only short vowels and consonants.[69] Masculine nouns are therefore "older" because they apparently contain vowel sounds that have existed in language since its origin. The vowels found in feminine nouns, on the other hand, date from a later period and were derived from the short, "masculine" vowels already in existence.

Words of the masculine gender are furthermore primary, according to Grimm, because of the way they behave when declined: "In its consonant strong ending, the masculine tends to retain the strict word form [. . .] the feminine characteristic is a weak A that also adheres in the accusative. The form of the feminine therefore appears as a milder form of the rougher masculine" (*DG* 313). Why the fact that masculine nouns retain their form during declension *means* that they preexisted feminine nouns, Grimm again does not explain. Instead, he offers eleven other points in favor of his thesis that "the masculine showed itself to be the liveliest, most powerful and most primary of all" (*DG* 313). These points include: that masculine nouns formally distinguish between nominative and accusative, whereas feminine nouns do not; that the genitive and dative forms of certain feminine pronouns and adjectives "appear to have been created by the genitive singular masculine"; that words derived from masculine pronouns, such as the German word *man*, are used for men as well as women; that feminine nouns can be derived from masculine words (*Herrin* from *Herr*), but rarely vice versa; and that "poetic can be derived from poet, gardening from gardener, but not poetess-ish from poetess, gardeness-ish from gardeness" (*DG* 314). He thus reaches the conclusion that "strict consonantism, short vowels and greater formative ability hereby determine the position of the masculine before the feminine, which has long countered those consonant vowels, those short vowels, and is more passive in nature" (*DG* 315).

My objective here is not to evaluate the veracity of Grimm's linguistic claims but to draw attention to the way that they are made to endorse a hierarchical conception of the sexes. It may be true that certain feminine nouns have long vowels, but Grimm's assertion that feminine nouns are "more passive in nature"—and his general association of femininity with passivity—relies on more than just linguistic data. In the section on the gender of inanimate nouns, for example, Grimm provides a list of weak feminine nouns derived from strong masculine ones. Referring to Genesis 2:22, or the creation of Eve from the rib of Adam, Grimm concludes: "Even if one doesn't find here any confirmation of the myth that woman was created out of man, woman's dependence on man can nonetheless be deduced from this inclination of language" (*DG* 348). Grimm presents this conclusion as purely the result of linguistic analysis, but his deductions in fact work both ways.

Buttressing his claim about language's "inclination" to generate feminine nouns from masculine ones is the assumed passivity of the female sex, which puts such linguistic arguments in line with the "natural" order.

The Dutch linguist Willem Gerard Brill makes a similar assertion in 1846, when he contends that the masculine gender precedes and "gives birth" to the feminine. According to Brill's theory, there was originally only one gender, what we now know as the masculine. The feminine noun class was created later, when language had lost its energetic power, and people began using softer forms of inflection—"for how could the idea of passivity, associated with certain subjects, be more appropriately presented than in the image of femininity?"[70] Brill's theory varies slightly from Grimm's in that he suggests that gender is completely a property of inflection (weak forms of inflection are retroactively assigned the feminine gender), whereas Grimm asserts a connection between the meaning of a word and its gender (words that designate perceived "weak" or "passive" objects, for instance, are feminine). Nonetheless, they share the same overarching position: namely, that the masculine is not only superior, but also antecedent, to the feminine. Such claims of masculine primacy are not restricted to linguistic discourse. The medical doctor Philipp Franz von Walther, for example, had claimed earlier in the century that the male sex is "the primordial, something by itself, in all its attributes, purely positive," while the female is "purely negative, [exists] only in opposition to the masculine, only through it and insofar as it gives it part of its essence."[71] Like Wilhelm von Humboldt's *Horen* texts from the 1790s, Walther argues that the process of conception is totally activated by the male sex. According to Walther's theory, man creates both himself and woman: "The power of the man creates itself and its counterpart in woman, and unites with it."[72] Underlying Walther's physiology is a fantasy of procreation that dispenses with the female sex altogether.

The masculine is similarly omnipotent in Grimm's work, where masculinity is a principle of household organization as well as an abstract concept. In the preface to the *Deutsches Wörterbuch*, written together with his brother Wilhelm, Grimm suggests an affinity between Sprachwissenschaft and religious faith, turning the study of the German language into a kind of devotional practice: "The dictionary could become a household necessity and be read with eagerness, often with devotion. Why shouldn't the father dig out a few words and, in the evening, use them to test his boys' language skills and at the same time refresh his own?"[73] Similar to Herder's notion of the *Familienfortbildung der Sprache*, which has man give language to woman, the Grimms imagine that the mother passively receives language from her husband; she "would gladly listen."[74] Whether one turns to the old source of truth (religion) or to the new (language), one lesson remains constant: man is primary, woman secondary and dependent.

In a recent study, Jakob Norberg has interpreted Jacob Grimm's theory of language pedagogy as a means of nation building, highlighting the gendered "iconography" that this project employed. "Nationalists of Grimm's era understood that the school served as an indispensable instrument of nation building," Norberg writes,

> and yet they preferred the image of the mother whispering to her child over the image of the schoolteacher instructing his pupils, for an honest recognition of mass schooling could suggest that the nation represented a willed political project rather than a natural, pre-political ground. [. . .] the recognition that nationhood was partially the outcome of large-scale schooling efforts would also sideline the figure of the philologist, whose political vocation depended on the importance of mediation between the natural community of the people and the ruling elite.[75]

Grimm solved this problem by turning the schoolteacher into a (metaphorical) wet nurse. The teacher was imagined as a mediating figure who, as Grimm writes, "like a wet-nurse [amme] holds the breast toward the infant, pours in the simple food of the first knowledge into the child, nourishes, prepares and instructs it in all things."[76] As in Kittler's examples of maternal instruction from earlier in the century, though, Grimm's mother does not have the authority to create or analyze language herself. She provides the material body through which language—and thereby the spirit of the nation—is perpetuated. The Grimms invoke a similar logic in a preface to their Kinder- und Hausmärchen when they describe the source of many of their stories, an old woman named Dorothea Viehmann who acts as a vehicle for cultural conservation: "She never changes anything when she retells a story and corrects mistakes as soon as she notices them even right in the middle of the telling."[77]

Returning now to Deutsche Grammatik, we can see that Grimm's conflation of grammatical and natural categories continues throughout the chapter on genus. Although his discussion of "masculine" and "feminine" vowels might suggest that he understands gender to be determined by a noun's form, Grimm ultimately returns to a semantic explanation, arguing that a noun's gender is best understood through its meaning. For the genders of animate objects (which possess what Grimm terms natural gender), this is relatively straightforward: Mann is a masculine noun because men are male. For inanimate objects (which possess grammatical gender), Grimm furnishes a theory of analogy. "The only valid or fertile [fruchtbar] way," he explains, "to present grammatical gender seems, to me, to be the one that takes the meaning of words into account; only in this way can one possibly succeed in detecting analogies that the human imagination has indulged by transferring natural sex onto an incalculable number of other nouns" (*DG* 358). In short, *Deutsche Grammatik* locates gender in the signified rather than the signifier. For Grimm, the relation between the signifier and the signified is not arbitrary. This is why masculine nouns apparently have short vowels and strong consonants, qualities that Grimm associates with masculinity. Yet it is important to emphasize that, according to Grimm's theory, masculine nouns are not masculine simply *because* they have "masculine" sounds or suffixes. Instead, they have these properties because they stand for something "masculine," an object or concept perceived to have an analogical relation to the male sex. "Masculine" form is motivated by "masculine" meaning. According to Grimm's theory of language, as Tuska Benes explains, in the original Germanic Ursprache, form and meaning were united as one; every object and experience had its perfect, corresponding mode of linguistic expression.[78] Grimm laments that we have fallen away from this original language, but also celebrates that traces still exist in modern Germanic tongues, which allows language to be a source of knowledge about the Germanic people and their history.[79]

Unlike the Enlightenment texts I discussed earlier, which contend that grammatical gender is based on caprice and therefore inaccessible to the modern speaker, *Deutsche Grammatik* takes the motivation behind grammatical gender to be fundamentally intelligible. The universal "truth" of sexual difference provides the code through which the modern scholar of language can decipher or detect the reasoning behind a word's grammatical gender. This is a practice in which Grimm engages repeatedly. He suggests (to provide just a few examples) that the nouns for many animals—when German does not distinguish between the male and female members of the species—can be explained by the animals' perceived "masculine" or "feminine" characteristics. *Maus* (mouse) is feminine because mice are small and fearful, while most birds are designated by feminine nouns because birds are small and graceful (*DG* 360). Nouns for *some* birds, however, are masculine because those birds are large and vicious (*DG* 361).

The naturalization of the division of the sexes, via language, is an overriding impulse of *Deutsche Grammatik*, which repeatedly turns the existence of grammatical gender into a confirmation of male dominance and authority. "Why are so many women's names formed with -*burg* [fortress] and -*gart* [garden]," asks a later edition of *Deutsche Grammatik*, "none with *haus* [house], *hof* [yard], *heim* [home]?"[80] This question suggests that words related to women should reflect women's "natural" place—the home, or pri-

vate sphere. Many scholars have analyzed Grimm's emphasis on male authority in his work on legal history and in his compilation of the *Kinder- und
Hausmärchen*. John Toews, for instance, notes that Jacob Grimm prefaces
his *Deutsche Mythologie* (1835) with an "emphasis on the patriarchal household arrangements of the ancient German gods and the patriarchal form of
ancient German law,"[81] while Orrin W. Robinson shows that one function
of pronouns in the fairy tales is to suggest the "expected or proper roles of
girls and boys in German society," leading to female characters being "less
individualized" than their male counterparts.[82] Similarly, Jack Zipes has argued that many of the Grimms' fairy tales support a process of male socialization that endorses "male domination" and "benevolent patriarchal rule."[83]
Deutsche Grammatik not only valorizes patriarchal authority but also takes
it as a methodology. When Grimm writes that a recourse to meaning is the
"only valid or *fertile* way to present grammatical gender," he portrays Sprachwissenschaft as a kind of sexual activity, where the fertile mind of the scientist
combines with the passive material of the archive to produce new knowledge
(*DG* 358).[84] Like the masculine gender itself, the male language scientist is
"agile," "creative," and "active," giving form to linguistic matter through his
categories and subcategories, his comparisons and derivations.

A Meaningless Form

Jacob Grimm died in 1863, but his *Deutsche Grammatik* lived on and was
republished in 1890 with a new preface by the philologist Gustav Roethe.
Grimm's chapter on *genus*, hailed by its editors as the "most significant" part
of *Deutsche Grammatik*, became a topic of debate beginning in the late
1880s—specifically, between Roethe and several members of the Neogrammarian school, including Karl Brugmann and Victor Michels.[85] A group of
linguists based in Leipzig who advocated a formalist approach to the study
of language, the Neogrammarians differed from earlier language scientists in
both their understanding of grammatical gender and their general approach
to their subject. Brugmann often referred to Grimm's method as "romantic"—
referring, perhaps, not only to literary romanticism, a movement with which
the Grimm brothers were tangentially associated, but also to the erotics of
Grimm's theory, which places sexual difference, and sexual copulation, at its
center.[86]

In attacking the theory presented in Grimm's *Deutsche Grammatik*, the
Neogrammarians were attacking a relatively entrenched understanding of
gender's origin and significance since, by the time of its republication in 1890,
Deutsche Grammatik had become a well-known text. Marcin Kilarski explains

that Grimm's work "constituted the basis for other 19th century accounts of the nature and origin of Indo-European gender," such as those of Franz Bopp, Wilhelm Bleek, and Karl Heyse; other prominent scholars, including August Friedrich Pott and Heymann Steinthal, also championed a similar understanding of the origin of grammatical gender.[87] In contrast to Grimm and the scholars following in his footsteps, the Neogrammarians argue that grammatical gender is internally rather than externally motivated—that is, noun genders develop based on a word's formal properties, not according to how the first speakers perceived the object it represents. In this way, they de-emphasize sexual difference as a necessary, formative power both in human life and in the development of language.

In 1880, for example, the Neogrammarian Hermann Paul argues in his *Prinzipien der Sprachgeschichte* (Principles of the History of Language) that "neither natural sex nor that ascribed by fancy has anything to do with gram-matical gender."[88] Instead, he suggests that grammatical gender can be ex-plained either as the result of a "stem ending" that at one point acquired a reference to a male or female individual or as "purely accidental."[89] Brugmann makes a similar argument nine years later, when he accuses Grimm of hav-ing "sexualized" language. Even if it were true that our "primitive" ancestors personified inanimate objects, Brugmann contends, this does not *necessarily* mean that they also assigned them a sex. In 1890, Roethe counters Brugmann with the following:

> If I personify something, that is, imagine an irrational or inanimate being as a person, as a human being, I have to imagine it as a man or a woman: per-sonification without a conception of sex is an absurdity [*personification ohne geschlechtsanschauung ist ein unding*]. It is completely impossible to think of looking at a sexless human; because "human" is just an indistinguishable abstraction of man or woman, with which the imagination cannot do any-thing. "The human," not "man" or "woman," is there only for an abstract way of thinking, which then has nothing to do with sensual imagination. This can be observed very precisely. The child who personifies his stick while playing does not say, "You are a person," but "You are a man," even better, "You are the papa." The deaf-mute, of which Degerando reports, saw in the moon "a sovereign woman," without being suspected of being influenced by the gram-matical gender of "la lune."[90]

This point of disagreement between Roethe and Brugmann may seem ar-cane, but the stakes of their dispute are in fact quite high. In the above quo-tation, Roethe is arguing for nothing less than the essential status of sexual difference: "Personification without a conception of sex is an absurdity." For Roethe, sex is a constitutive quality of human existence. There is no general conception of the human that is not already sexed, since "human" is just "an unvisualizable abstraction of man or woman." Implied here—and in fact articulated explicitly by Grimm—is the assumption that man is the universal, woman the derivative.

Conversely, what Brugmann suggests in his essays on grammatical gender is a conception of the human, and of language, that is independent of sex. In his 1890 response to Roethe, Brugmann again contends that personhood does not require "sexualization":

> Of course, I did not deny that our ancestors, who lived in primitive cultural conditions, had a more active instinct for personification than we do today. I have also admitted that it was in principle conceivable that in language, to the greatest extent possible, lifeless things were treated as something living, personal. I only declared it improbable that every object and concept that was personified was also in a certain way sexualized, seen as either a male or female being.[91]

Brugmann's claim that it is possible to have an idea of the human that is not already sexed comprises, compared to the other texts on gender that I have discussed, a significant deemphasis of the role of sex in human identity. Brugmann's critique of Grimm and Roethe contains an unwitting critique of the dominant nineteenth-century ideology of sexual difference, which posits that sex orders and articulates the major aspects of life.

In addition to offering a new perspective on the significance of sexual difference, Brugmann also, of course, offers a new theory of the origin of grammatical gender. Brugmann's explanation of gender proceeds as follows: in Indo-European, the suffixes -ā and -iē became associated with the feminine gender because a small group of words with these suffixes referred to women.[92] These suffixes then become productive such that other words with the same suffixes were also associated with the feminine gender, even though these words had no semantic relation to women. Later, Brugmann revises this thesis and argues that the suffixes -ā and -iē originally did not even refer to women at all, but rather to "abstracts and collectives." They gained a feminine association only later, through a series of derivations that can be observed, for example, with the German word *Huhn*: "This meant at first the cocks and the hens together, then the flock of female fowl, and finally the individual female fowl." Thus, Brugmann continues, "If the suffixes–ā- and–iē- implanted themselves in this manner in a number of words of feminine signification, the idea of feminine sex could attach itself to the suffixes, and they could acquire this additional shade of meaning."[93]

Although Brugmann's theory establishes an initial connection between grammatical gender and biological sex, this connection is quickly severed. First, for Brugmann, the fact that *these particular suffixes* are associated with the feminine gender is completely arbitrary. There is no intrinsic relation between the sounds -*ā* and -*iē* and any concept of maternity or femininity. This runs counter to Grimm's theory of gendered phonology, which suggests that consonants and short vowels are "masculine" because they bear a resemblance to characteristics like strength, power, and activity. Second, in Brugmann's theory, once the suffixes become productive and the feminine categorization is applied to other nouns, the feminine gender ceases to hold any relation to the female sex, and gender becomes purely a means of formal classification. Nonetheless, even this theory ultimately cannot be corroborated, as Brugmann admits to an American lecture audience, while managing one last jab at his adversaries:

> The solution which I have presented to you can unfortunately never be absolutely proved; for we have to do with a period in the history of language in which we cannot go a step further than simple hypothesis. It can be said, however, of our explanation, and it is indeed its strongest claim over the theory of Adelung and Grimm, that it keeps within the limits of phenomena which are among the best substantiated in the history of the Indo-European language family, and which may be observed in the very latest phases of its development.[94]

Whereas Grimm asserted a reciprocal relationship between language and the human subject, using human and linguistic categories to decipher each other, Brugmann argues that language is a completely autonomous system, interpretable only within itself and on its own terms.

Neither Brugmann's nor Grimm's theory is accepted by linguists today. The question of grammatical gender's origins, moreover, which must necessarily be answered through conjecture, is rarely recognized as a legitimate topic for the study of language in the twenty-first century. So why read these texts? For one thing, they reveal something about how and where sexual difference was debated historically. As we have seen, in texts on grammatical gender, scholars debated not only the nature of language but also the nature of the sexes, the difference between them, and the meaning of that difference. The history of language science cannot be told accurately if we do not acknowledge the role that ideas about the sexes have played in the formation of this discipline and the conceptualization of its subject.

Moreover, the texts of Grimm and Brugmann show how the subjection of women is accomplished intellectually. The nineteenth century was a period in which the nature of sexual difference—and, correspondingly, the social and political status of women—was under intense scrutiny. Theories of grammatical gender were written during the same era in which moral physiologists were finding copious support for the natural inferiority of the female mind and body; in which an ideology of "separate spheres" increasingly excluded women from public and intellectual life; and in which women's movements campaigned for advances in social rights and education. German texts on language not only reflected these debates over sexual difference but also provided a rationale for the idea that the division of the sexes is natural and universal. As theories of grammatical gender established the importance of sex for language itself, they also generated a language *of* sex, a way to construct and evidence the differences between women and men.

4

Women Writing on Language

Judith Shakespeare, sister to William and product of Virginia Woolf's prodigious imagination, had "the quickest fancy, a gift like her brother's, for the tune of words. Like him, she had a taste for the theatre."[1] As Woolf envisions it in *A Room of One's Own*, however, Judith's "genius for fiction" is met with none of the opportunities that alight on her brother. After an attempt to join the theater in London proves futile, she ends up pregnant, committing suicide, and condemned to obscurity, "buried at some cross-roads where the omnibuses now stop outside the Elephant and Castle."[2]

Judith Shakespeare is a fictional character; for Woolf, that is precisely the point. Even if such a figure had existed, we would be unlikely to know about her. Like many real women, Judith's talents are snuffed out by her material circumstances and the gendered expectations of her era. If Woolf's "Judith" had miraculously been able to write, and then, even more miraculously, to have her work performed, she would have been unlikely to receive the same canonization as her brother. In this way, *A Room of One's Own* gives voice to a methodological challenge that besets anyone who would investigate historical works by women and other marginalized figures. How do you disinter an unmarked grave? How do you look for an absence?

*

Clementine von der Gabelentz, sister to eventual language scientist (*Sprachwissenschaftler*) Georg and daughter of linguist Hans Conon, also had a talent for languages. Along with her siblings, she profited from her father's instruction and vast library, which included "one of the largest collections of grammars, dictionaries and linguistic treatises" in the nineteenth-century world.[3] At the age of eighteen, Clementine put her training and resources to use, pro-

ducing a manuscript grammar of Lepcha, a language of the eastern Himala-
yas.[4] But whereas her brother Georg would go on to study at the universities
of Jena, Leipzig, and Dresden, becoming a regarded scholar of Chinese and
theorist of language science, the Lepcha grammar marked the end of Clementine's formal linguistic output. In 1873, at the age of twenty-four, she married the baron Börries von Münchhausen, and they took up residence in the
family castle.[5] As a married woman, Clementine worked in the textile arts,
producing needlepoints inspired by the materials she encountered on her
travels.[6] She also wrote biographies of Georg and another brother, Albert.

Would Clementine von Münchhausen have become a linguist, had she not
been born a woman? Would she even have wanted to? These are, of course,
unanswerable questions. Yet it is hard not to see in her considerable needle-
point collection the traces of a sublimated intellect and an interest in the same
themes that animated her father and brother's academic pursuits, such as the
diversity of cultural traditions and their artifacts. We know from a biography
of Clementine written by her husband, as well as her own account of Georg's
life, that she was not only interested in Sprachwissenschaft but also possessed
an aptitude for it. By the time she left her father's protection for her husband's,
she had developed working knowledge of over a dozen languages, including
Finnish, Estonian, Norwegian, Latin, and Lithuanian, and frequently undertook difficult projects in translation and linguistic analysis.[7]

Her brother also relied on her help to produce his published texts. When
writing his *Ideen zu einer vergleichenden Syntax* (Ideas on comparative
syntax)—a work designated "epoch-making" by Heymann Steinthal—Georg
had neither the time nor the resources to complete the necessary research.
Enter Clementine, who was tasked with investigating the position of subject
and predicate in Ancient Egyptian.[8] Years later, when Georg's book *Die Sprachwissenschaft* is in press, he writes to Clementine that he will come to her with
the printed sheets, as he has done before: "You are an appreciative reader, and
with my books I feel a little like Heine with his verses, I dream of angels who
sit around me and praise me. You are one of them!"[9]

As his muse, Clementine's relationship to Georg is symbolically reconfig-
ured from sibling to wife. When Georg would return to their parents' home
for a visit, Clementine writes, she would run to embrace him. "The bridal
couple!" someone would inevitably chuckle.[10] Georg then "occasionally said
in all seriousness that he found sibling marriage, as was customary in ancient
Egyptian royal families, very lovely, and would have taken me straight away."[11]
Clementine's text implicitly contrasts her own relationship to Georg with that
of his first wife, whom he eventually divorced. Whereas Clementine provides
Georg with the support necessary to produce his language-scientific insights,

his first wife was, in Clementine's telling, "admittedly clever and urbane, but absolutely unscientific and talentless, and could never have appreciated his scientific importance."[12] A second biographical sketch written by Clementine further emphasizes the extent to which Georg's first wife held him back in his studies.[13] The role of woman, her descriptions suggest, is that of helpmate, supporting and facilitating male genius.

In the texts of Clementine von Münchhausen née Gabelentz, the theories of Herder et al. play out in real time. As Clementine writes, Sprachwissenschaft is Georg's patrimony: Hans Conon "passed down to him [*vererbt*]" his linguistic interests,[14] and this is an inheritance that Georg proudly gives to his own son.[15] Thus does his mother, Adolfine, complain that whenever Georg returns home, he "absorbs Papa completely."[16] But how could it be otherwise? There is no room for female figures in the patriarchal perpetuation of language and language study. Masculine scientific analysis usurps and displaces the feminine domestic realm, even as the domestic realm is what makes this science possible. According to Clementine's text, when Adolfine gives Hans Conon a manuscript as a gift—ignorantly mistaking Ethiopian for Coptic!—she grumbles that he spends the next eight days so absorbed in his work that he does not address anyone in the household. To paraphrase Sara Ahmed's description of Husserl's writing desk: "The family home is only ever co-perceived, and allows the linguist to do his work."[17] Indeed, in the "refuge" of the family home, the linguist should be regarded as a "priest in his sanctuary," argues a biography of the linguist August Schleicher.[18] Incidentally, the same biographer records that Schleicher was an "enemy of women" (*Weiberfeind*) until he "got a good, pleasing girl from his homeland"[19]—a girl whose goodness, one presumes, had as much to do with her family wealth as with her willingness to serve as his uncredited scholarly contributor.[20]

As the Schleicher and Gabelentz biographies make clear, tropes such as the masculinized subject of language invention, or language as patrimony, structure not just philosophical and scientific texts but also lives—as they were lived and as they are remembered. The same ideas about women's limited capacities that facilitated their theoretical-philosophical marginalization also facilitated their *actual* exclusion from the language disciplines. Clementine von Münchhausen has been memorialized as a practitioner of women's arts and the matriarch of an aristocratic family. Her thinking about languages and their histories exists only in the traces she left on her brother's oeuvre, and we can but speculate about what she might have produced, had she been afforded the opportunity. Such speculation is not without merit. Counterfactual history, as Virginia Woolf shows, can be an effective method of critique. Yet the counterfactual, as a methodology, also has its limits, the most obvious being that it restricts feminist criticism to the realm of fantasy. This tension—between the desire to supplement the archive with imagined material and to expose what it lacks, and why—is encapsulated in Saidiya Hartman's term *critical fabulation*. In Hartman's work on the histories of enslaved women, "critical fabulation" refers to a practice of "straining against the limits of the archive to write a cultural history of the captive, and, at the same time, enacting the impossibility of representing the lives of the captives precisely through the process of narration."[21]

If we try to turn from the counterfactual to the factual, and from imaginary texts by women to those they actually produced, we face a dearth of sources. The paucity of language-philosophical and language-scientific texts by women is a direct result of their exclusion from these disciplines. In most areas of Germany and Austria, women gained the right to matriculate at university only around 1900, and it would be longer still until social norms caught up with legal rights. In Prussia, even once women were allowed to attend university in 1908, faculty members were entitled to "exclude matriculated women students from their classes," a prerogative of which Jacob Grimm's inheritor, Gustav Roethe, availed himself quite frequently.[22] Women were rarely welcomed in the institutional spaces where ideas about language and its analysis were being worked out. This included not only the university lecture hall but also scientific and philosophical journals. Friedrich Techmer's influential *Internationale Zeitschrift für allgemeine Sprachwissenschaft* (International journal of general linguistics), for instance, counted no women on its advisory board or among its contributors for the entirety of its existence between 1884 and 1890.[23]

To investigate women's thinking about language prior to their participation in the relevant institutions and disciplines, then, we have several options: (1) discuss it as an absence; (2) highlight women's influences on men's published works; and (3) study other kinds of works by women, such as literature, translations, or textile arts, and attempt to extrapolate from them an argument about language. These approaches have been employed in a variety of ways. The editors' introduction to *Women in the History of Linguistics*, for instance, engages the first approach when it argues that, in a number of nations, women did not take part in linguistic research until the twentieth century.[24] Sabrina Ebbersmeyer does something similar for philosophy, showing how, beginning around 1800, the German historiography of philosophy excluded even ancient female philosophers from its accounts, correlating with the strict exclusion of women from the discipline of philosophy itself in the nineteenth century.[25] Nicola McLelland's study of women in the German linguistic tradition primarily employs the second and third approaches, as she highlights "female translators and poets as scholars of the German language" in the seventeenth and eighteenth centuries, and emphasizes Caroline von Humboldt and Dorothea Grimm's contributions to their husbands' work.[26] Alan Corkhill, on the other hand, reads women's letters and literary texts to make a more tenuous argument for "female language theory in the age of Goethe."[27] In recent work on women in the history of philosophy, Sarah Tyson suggests a new method of what she calls "transformative reclamation," which aims to make the discipline of philosophy more inclusive by expanding its archive. And yet, insofar as her approach encourages "more freedom in speculating from historical texts to new contemporary possibilities," it exchanges a careful study of the past for a theorization of the present.[28]

The strained results that such methodologies can, at times, produce is less the fault of the scholars who undertake them than of the historical records they investigate. All recuperative approaches are inevitably unsatisfying in one way or another, as the historical fact of women's inequality poses an irresolvable challenge to feminist criticism. No method can uncover what was never written, nor can it transfigure what was written into a genre that women were discouraged from producing. The following chapter offers no grand solution to this aporia of feminist historiography, only an acknowledgment of it, and a commitment to analyzing the texts I study on their own terms.

Focusing on women-authored texts, this chapter investigates three counterexamples to the works examined in the book so far, where tropes of woman and femininity suggested that women are ill-suited to the creation and conceptualization of language. Chapter 4 spans a later chronology than previous chapters because, to put it bluntly, this is where the texts are. As discussed above, women did not begin to participate formally in the academic disciplines of linguistics, philology, and philosophy until close to the turn of the twentieth century. I therefore focus on three figures who wrote during this period: Carla Wenckebach, Carolina Michaëlis de Vasconcellos, and Elise Richter. Each of these thinkers employs a different approach to constructing textual authority in relation to gendered identity—an issue that they are forced to confront not only in terms of the content of their scholarly works but in their own linguistic production as well. As they stage their voices in relation to masculine academic discourse, they participate in contemporaneous debates about women's speech and its capacity for scientific analysis.

Carla Wenckebach's Universal Subjects

Cató Wenckebach, born in Germany in 1853, was by all accounts a sharp intellect and a gifted student of languages. While her successful career as a college lecturer was unusual for a woman of her time, her subject—foreign-language instruction—was a common field for women to enter.[29] As a young woman, Wenckebach studied at a women's teaching college in northern Germany, worked as a governess in Scotland and Russia, and eventually immigrated to the United States, where she attempted to fashion herself as a journalist. When this career stalled, she undertook further study in foreign-language pedagogy and secured a position teaching German at Wellesley College in 1883. In her time at Wellesley, Wenckebach published an impressive number of pedagogical works, focusing especially on teaching German to American students. Her texts advocate for a "natural" method of language instruction informed by the theories of Swiss pedagogue Johann Heinrich Pestalozzi, among others.[30] In this way, Wenckebach professionalized the role of maternal language pedagogy that Friedrich Kittler describes, in which the mother acts as the conduit through which the (male) child learns to speak.[31]

According to a biography written by her Wellesley colleague Margarethe Müller, however, Wenckebach eschewed normative femininity.[32] For Müller, Wenckebach represents an increasingly prominent type of woman, the woman "whose instincts and interests are intellectual rather than domestic."[33] This woman requires her own origin narrative, and what is particularly interesting about Müller's biography is that it constructs this narrative by activating the same tropes of familial language development favored by the likes of Herder and Fichte, albeit with one significant reversal.

Wenckebach flouted gendered expectations already as a young child, Müller writes, from her "utter contempt" for needlework and knitting, to her masculine clothes and male friends,[34] to the taste for cigars that she developed at the alarmingly early age of four.[35] In Müller's text, Wenckebach's "masculine" preferences are the result of her identification with her father, who molds his daughter in his own image. Although Wenckebach père also has a son, Claus, he is—Müller suggests—more of a girl than a boy: a "dainty" child who "wept when he got his hands or his clothes soiled, and who could play for hours making finery for his dolls."[36] Thus Wenckebach must be the one to bear her father's legacy. Notably, this patrimonial transmission begins as her father teaches her his regionally inflected language. Sitting on her father's lap, Wenckebach would "[repeat] the big words that he rolled out for her, words, maybe, like *Podbielski, Sebastopol, Unabhängigkeitskampf,* and others that were in the air at that time," words which were "as interesting to [her] as his dolls were to Claus."[37] When, in America, Wenckebach changes her first name to "Carla" in honor of her father, Carl, we can read this as the culmination of the process of paternal identification initiated in her youth.[38]

Yet Wenckebach's nonconformism only goes so far. Her professional

publications strictly adhere to conventional conceptions of masculinity and
femininity and place her on the "feminine" side of a gender binary that she
so often flouted in life. In this way, Wenckebach's texts demonstrate the gender rigidity of the nineteenth-century discourse on language. The practice
sentences in her *Deutsche Grammatik für Amerikaner* (German grammar for
Americans), for example, which was first published in 1884, are premised
upon a traditional division of gendered characteristics and labor. To teach adjective endings, the text declines "der starke Mann" (the strong man) through
various cases. *Strong*, however, apparently cannot function as an representa-
tive adjective for the feminine, and the *Grammatik* switches to a new descrip-
tor, "die junge Mutter" (the young mother), in the subsequent section.[39] And
while elsewhere in the text, a female subject "spinnt" (spins textiles)—ironic,
given, Wenckebach's alleged hatred for such "womanly" arts—the *Grammar*
exemplifies a male subject in the same section through a student who learns
diligently and collects a prize.[40] Significant, too, is the way that Wenckebach's
practice sentences place fathers in relation to their sons much more often
than their daughters.[41]

In 1886, Wenckebach published, with her sister, a supplementary handbook to the *Grammar*.[42] This handbook, they write, "is especially designed for
the use of the beginning and middle classes in German at Wellesley College,"[43]
an all-girls school. Yet the Wenckebachs use masculine pronouns when they
refer to exemplary students and teachers. The same occurs in their textbook
published one year later, the *Deutsches Lesebuch* (German primer),[44] whose
lessons they also describe as being based on their experience at Wellesley. The
discussions of "the teacher" and "the student" are written in English, where
the nouns do not have grammatical gender; that is to say, masculine pronouns are only conventionally, not grammatically, necessitated.[45] It is telling
that a book written by two women—for use in and based on the instruction
of women students—so disregards its context of production that it portrays
its subjects in terms of a male standard. It suggests the difficulty of imagining
and articulating a universal subject outside of the masculine frame.

Given the way that they masculinize language teachers and learners, it is
not surprising that the Wenckebachs stage their own text as a handmaiden
of research rather than research itself: this is the only role that the feminine
may inhabit. In the preface to the *Lesebuch*, they write that the knowledge of
"Arian and Germanic languages" that their book facilitates should arouse "in
the student a genuine interest in the scientifie [sic] study of language, and
mak[e] it possible for him later to solve successfully more difficult problems
in the department of philology."[46] The function of their text, in other words,
is to produce the linguists of the future.

Are the Wenckebachs' female students subject to the same pedagogical formation as the text's exemplary male learner? Women at Wellesley presumably encountered lessons identical to those published in the *Deutsches Lesebuch*, which not only includes exercises in German vocabulary and grammar but also introductory lectures on the history of the German language and on Grimm's law of sound change. Even the Wenckebachs' emphasis on pronunciation should, by their own telling, incite future academic pursuits: "A thorough knowledge of the sounds is a great acquisition; it is indeed the preliminary step in all intelligent philological study."[47] But within these lessons about linguistic production are also, as we have seen, implicit lessons about gender norms and the divergent activities suitable for the two sexes. The Wenckebachs exhibit a tension between the content of their work and what that work performs. In the Wellesley classroom, they teach women German in a way meant to foster higher study; yet their lessons limit the possibilities of how women may imagine themselves and their achievements.

The Self-Creations of Carolina Michaëlis de Vasconcellos

Whereas the texts of Carla Wenckebach adhere to and even perpetuate conventional gender norms, the early work of Carolina Michaëlis de Vasconcellos questions the gendered assumptions employed by her predecessors. *Studien zur romanischen Wortschöpfung* (Studies on Romance word formation), which Michaëlis de Vasconcellos published in 1876, opens with a problem. The author has been scooped, and by the "master of Romance philology," no less. The renowned Romanist Friedrich Diez had issued his *Romanische Wortschöpfung* (Romance word formation) the year before, when her manuscript was already completed. In the opening pages, she acknowledges this awkward overlap and performs the expected pieties of feminine submission. Even though it is three hundred pages long, she insists that her text is not a book but a "booklet," a "little work," a "patchwork."[48] Thus, she explains, she has added the phrase "studies on" to the title in order to dispel any claims to comprehensive knowledge. Twice she requests the harsh judgment that she knows the work will merit, given its many faults.[49] And twice she recognizes her subordinate position as a woman working in a discipline predominated by men, hoping that her readers will not spare her their sharpest criticism "for example out of consideration for feminine tenderness."[50]

Were this the extent of the *Studien*'s metatextual reflections, it would appeal to us primarily as evidence of the historical restrictions women faced in constructing an authorial voice. What is remarkable about Michaëlis de Vasconcellos's study, however, is how it simultaneously reproduces and undermines the gendered hierarchies of her era. By identifying with and defending her subject, the Romance languages, her text loosens language-philosophical concepts from the gendered connotations in which they had become entrenched and reveals metaphorical "daughters" to be richer and more complex than the fathers that produced them.

The Romance languages suffer from the same negative reputation that Michaëlis de Vasconcellos initially affords her own text: deficient and feminine. Romance languages, regarded as "no more than one of Latin's daughters," are cast as impoverished maidens in need of chivalric heroics, she writes.[51] Specialized scholars rush to their defense "as their knights to lodge a lance for them."[52] Michaëlis de Vasconcellos is not alone in characterizing Latin as masculine and the Romance languages as feminine, or what one mid-nineteenth-century linguist referred to as the "daughters of the world-ruling Latin."[53] Latin and Greek were associated with masculinity because they were "stronger" and closer to the original language created by man, whereas languages like French and Spanish were, like women themselves, merely derivative. In the nineteenth century, the study of ancient languages was thus considered "neither feasible nor suitable for women," writes James Albisetti.[54] Knowledge of Greek and Latin was irrelevant to women's domestic duties; moreover, the "logical activity" effected by learning Latin was considered, as one pedagogue put it in the 1830s, "not exactly feminine."[55] As this pedagogue imagines it, women who learn the wrong languages threaten the stability of the domestic sphere: "A woman who is educated in the classical languages—that is, a learned woman—is seldom a happy one for man and the household."[56] If German women want to learn a foreign language, he suggests, let them try Italian!

Although Michaëlis de Vasconcellos does not explicitly contest the feminization of the Romance languages, she argues against their subordination to "masculine" Latin. Rising from the rubble of the collapsing Roman world, she writes, each Romance language has come to constitute its own domain, "different but no worse" than Latin.[57] In fact, Michaëlis de Vasconcellos asserts that the Romance languages should be recognized for what she calls, variously, their "self-direction/activity" (Selbsttätigkeit) and "own self-creating activity" (eigene selbstschaffende Tätigkeit).[58] As we have seen in the theories of Fichte, Humboldt, and Grimm, among others, Selbsttätigkeit is linked to masculinity in this period. In the Studien zur romanischen Wortschöpfung, Michaëlis de Vasconcellos references and even quotes from Jacob Grimm; she was undoubtedly familiar with the gendered concepts produced by his system of thought and his association of Selbsttätigkeit with masculinity. In claiming self-creation for the "poor" and "feminine" Romance languages, then, she undermines the strict gender binary that structured so many language-philosophical systems of the era—and also, implicitly, claims this capacity for herself.

The *Studien* mirrors the Romance languages not only because it is ostensibly "feminine" and deficient, but also because it conforms to the same model of development. According to Michaëlis de Vasconcellos, it is the nature of all languages to progress and improve. Later-developed languages are always "richer" than their predecessors—thus Spanish, Italian, and French are richer than Latin.[59] Couched in a self-deprecating paragraph is a similar description of her own research. Her work exists in a constant state of improvement: "I myself continue to collect tirelessly, and since only the subject and its success are important to me, I will of course receive every expansion and every correction with joy and sincere thanks."[60]

In this way, Michaëlis de Vasconcellos replicates the structure of mutual identification that made possible the "congenial" resonance between Sprache and Sprachwissenschaftler in Humboldt, for example. But whereas in Humboldt, this resonance results from a shared masculine form, in the *Studien*, the genders of both subject and author are not so fixed. Michaëlis de Vasconcellos may be a "daughter" of Sprachwissenschaft, genuflecting to the requisite patriarchs, but her work is also—by her own logic—more complex and sophisticated than that of previous generations. In an era where there was no easy path for women to enter academia, Carolina Michaëlis de Vasconcellos, through "self-creating activity," forged her own way. Much of her training came from an autodidacticism enhanced by correspondence with other Romanists.[61] Despite ending her formal schooling only at the level of a *höhere Töchterschule*, an educational institution focused mainly on training women for domestic duties,[62] she went on to receive honorary doctorates from several universities, hold a professorship in Portugal, and author multiple works in linguistics.[63] When she writes in the *Studien* that her focus has been "the independence [Selbstätigkeit] of the Romance languages, the way in which they broke away from Latin in order to pursue their own paths and to enrich themselves,"[64] she could equally be describing her own trajectory. Just as her *Studien zur romanischen Wortschöpfung* untethers key concepts of language science from the genders to which they had been fettered, Michaëlis de Vasconcellos's career expanded the possibilities for women's institutional participation in linguistics.

Elise Richter's Androgynous Diagrams

One of the first women to matriculate at the university of Vienna, Elise Richter received not only a doctorate but also the more advanced qualification of

Habilitation. She went on to become the first woman to hold a university lec-
tureship in Austria, where she produced influential linguistic research.[65] This
remarkable life was ended by Hitler's annexation of Austria; in 1942, Richter
was murdered in Theresienstadt.

Born into a bourgeois, assimilated-Jewish family in Vienna in 1865, Richter was initially tutored at home with her sister. Although she studied English
and French as a child, it was only at age fourteen, while reading Friedrich
Mommsen's multivolume history of Rome (which included linguistic comparisons), that she learned of the methodologies of Sprachwissenschaft and
decided it would be her discipline. Yet her request for a textbook in Indo-
Germanic or Latin "was not fulfilled on the basis of its being 'unladylike' and
'mad,'" Richter writes in an autobiographical text. She and her sister, she thus
concludes, "hated our sex."[66]

Despite such disincentives, Richter persisted in her studies and in 1897
passed the *Matura* exam, which granted women entry to the university,
thanks to a new law.[67] Notably, Richter concentrated on Romance languages
in her doctoral work, although she also desired to study Sanskrit and Ancient
Greek—for which she certainly possessed the aptitude, as she had studied
Greek for the *Matura.*[68] In her autobiographical writings, Richter does not explain her decision for *Romanistik,* but it conforms to the notion that ancient
languages were an unsuitable subject for women.

Richter did not affiliate or identify with the feminist movements of the
early twentieth century, although this was, as Michaela Raggam-Blesch suggests, at least partially a strategic choice; "otherwise her academic career,
which still confronted numerous obstacles, would hardly have been possible"
in the male-dominated university.[69] As a scholar, she was prolific and wide-
ranging. Her published works cover general linguistics, Romance historical
linguistics, and phonetics.[70] Her scholarship on general and historical linguistics gives no mention of sex or gender, as is also the case with contemporaneous texts by men. By 1900, the gendered tropes and concepts that animated
so much of the previous century's conceptualization of language were no longer prevalent. Linguistics as a discipline had moved away from the kind of
speculative theorizing that accompanied philological analysis in Grimm and
Humboldt in order to focus only on what could be empirically described and
analyzed (cf. the Neogrammarians). Indeed, the origin of language—which
required conjecturing about the first humans and their conditions, and one
site where gendered descriptions of the human often appeared—was no longer considered a relevant subject of inquiry.[71]

Whereas a demonstrably male subject recedes from historical and general
linguistics around 1900, it persists in phonetics. The modern field of pho-

netics was formed in the nineteenth century, although the study of sound has likely existed ever since humans began investigating language, as E. F. Konrad Koerner observes.[72] Operating at the borders of linguistics, physiology, and physics, nineteenth-century phonetics included historical, theoretical, and experimental branches. Using diverse technological instruments such as the resonator, tinfoil cylinders, and the phonoautograph, experimental phoneticians attempted to measure and even reproduce the sounds of human language.[73] These technologies, writes Tobias Wilke, enabled researchers to "approach speech in a manner equivalent to other recordable bodily processes such as breathing, heartbeat, blood pressure, and pulse" and to "explore the physicality of speech—conceptually and on paper—by intertwining body and voice with various forms of phonetic visualization."[74] Such phonetic visualizations are of particular interest to me because they often represent their findings through diagrams. In portraying exemplary humans, texts endow their figures with male secondary sex characteristics—like facial hair or a prominent Adam's apple—which, although irrelevant to the matters of sound that they discuss, reveal their imagination of the human as necessarily sexed.

The sheer number of diagrams inserted into texts on phonetics in this period betrays a kind of mania for visualization. Images were marketed as a selling point by publishers. The 1914 *Einführung in die angewandte Phonetik* (Introduction to applied phonetics), for instance, by G. Panconcelli-Calzia, who ran an experimental phonetics laboratory at the University of Hamburg, boasts 118 figures and three collotype prints among its 131 total pages. It was common for phonetics books to borrow illustrations from one another, and here Richter's texts are no exception. Her two major works that discuss phonetics, *Wie wir sprechen* (How we speak; 1912) and *Lautbildungskunde: Einführung in die Phonetik* (Study of speech formation: Introduction to phonetics; 1922), adapt visual material from earlier texts, especially Friedrich Techmer's 1880 *Phonetik: Zur vergleichenden Physiologie der Stimme und Sprache* (Phonetics: On the comparative physiology of voice and language). Techmer's *Phonetik*—along with other foundational studies that Richter draws on, including Wilhelm Viëtor's *Elemente der Phonetik des Deutschen, Englischen und Französischen* (Elements of the phonetics of German, English and French; 1884), Hermann Gutzmann's *Physiologie der Stimme und Sprache* (Physiology of voice and language; 1909), and Panconcelli-Calzia's *Einführung*—illustrate their claims through a combination of sex-neutral and explicitly male human figures.[75] For instance, in Techmer's depictions of a laryngoscopy and rhinoscopy (which he actually borrows from another text), the scientist as well as the speaker are presented as male (figures 1 and 2).[76]

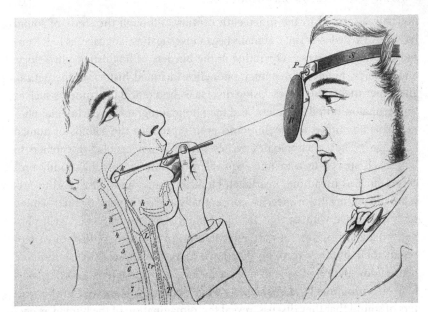

FIGURE 1. "Laryngoscopal examination of another person." Techmer, *Phonetik*, vol. 2, 30.

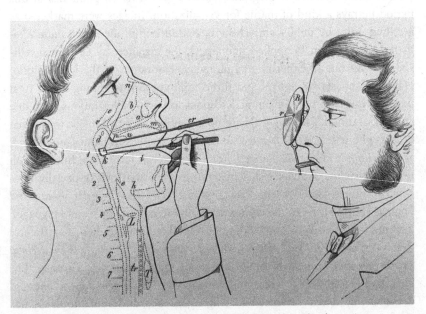

FIGURE 2. "Rhinoscopal examination of another person." Techmer, *Phonetik*, vol. 2, 35.

While both people appear masculine, this is most noticeable in the scientist, who is rendered with bushy sideburns, masculine clothing, and a short haircut. The examinations are not only images of intimacy—the figures are mere inches away, face to face—but also of penetration. It is no surprise, then, that the scientist is assigned the most obviously male characteristics. In the logic of heterosexual complementarity, penetration is the domain of the "active," virile male.

Similar male secondary sex characteristics are found throughout Techmer's text. The illustration of a palatal view of the production of the *i* sound, for example, has an unmistakable mustache (figure 3). When this *i* diagram appears in Richter's 1912 *Wie wir sprechen* (How we speak),[77] however, it has lost its facial hair (figure 4).

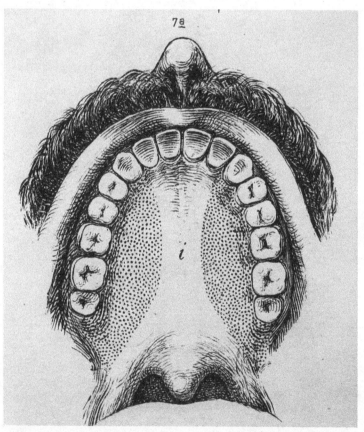

FIGURE 3. "Stomatoscope image of palatolingual articulations." Techmer, *Phonetik*, vol. 1, table III (appendix).

The same unmustachioed figure appears again in Richter's later text *Laut-bildungskunde*. There is no mistaking its derivation, since Richter cites Techmer as the source for this diagram. While men's grooming trends certainly changed between 1880 and 1912, the mustache remained in vogue in Europe through the First World War.[78] Richter's transformation of this diagram seems less about the mercuriality of fashion than about expanding the image's representational capabilities: reimagining the speaking subject as sex and gender neutral.

In her language-scientific texts, Richter never refers to her status as a woman. The same genderless persona that she cultivates as an author is reflected in her diagrams, which (except for one photograph, copied from elsewhere) betray no signs of sex or gender.[79] Indeed, in the only diagram that is original to *Wie wir sprechen*, the human appears in a disassembled, androgynous form (figure 5).

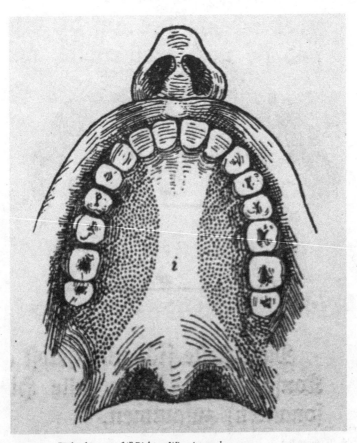

FIGURE 4. "Palatal image of *i*." Richter, *Wie wir sprechen*, 17.

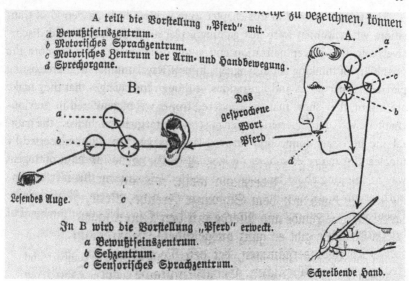

A teilt die Vorstellung „Pferd" mit.
~~~~~~~~~~ zu bezeichnen, können

a Bewußtseinszentrum.
b Motorisches Sprachzentrum.
c Motorisches Zentrum der Arm- und Handbewegung.
d Sprechorgane.

A.

B.

Das
gesprochene
Wort
Pferd

Lesendes Auge.

In B wird die Vorstellung „Pferd" erwedt.
a Bewußtseinszentrum.
b Sehzentrum.
c Sensorisches Sprachzentrum.

Schreibende Hand.

FIGURE 5. "Schematic representation of the communication process." Richter, *Wie wir sprechen*, 43.

The speaking head in this diagram, like the disembodied hand, cannot definitively be read as belonging to a man or a woman. In contrast to someone like Gustav Roethe, who, as we saw in chapter 3, contends that any image of the human is always already sexed, Richter's texts put forward a vision of the human that is neither male nor female. In the introduction to *Wie wir sprechen*, she writes that "we cannot imagine our human existence without language."[80] Whereas in the nineteenth century, the "we" of Sprachwissenschaft was a pronoun of exclusion—referring only to a collective of male scholars—Richter's text imagines a community of speakers and researchers beyond the gender binary. Sprachwissenschaft, Richter writes, the "exploration of the world with everything that is in it, everything earthly and everything spiritual, is the form and content of human experience."[81] This "everything" includes her as well.

✳

In the 1910s and 1920s, the number of women entering language disciplines at the university level grew, as did, correspondingly, the number of linguistic texts they published.[82] In the 1930s, despite Nazi antifeminist ideology, women continued to study in linguistic and philosophical disciplines, although their numbers dropped.[83] The postwar period then saw a gradual rise in women's participation in these fields, bolstered by the feminist movements of the later twentieth century.

The texts examined in this chapter were written during a period of transition, when women were just gaining a place—institutionally and discursively—in the conceptualization and study of language. They explore the possibility of thinking about language in new ways, unmoored to preexisting patriarchal narratives and figurations. Although the changes that they make are minor, they show that the gendered tropes we encountered in previous chapters were not impervious to change and reimagination. Indeed, the natural and necessary connection between language and masculinity asserted by nineteenth-century discourses on language is but one historically contingent way of thinking about language and gender. Recognizing this fact has import for our understanding of the past as well as the present. As Michèle Le Doeuf writes about one of John Locke's anecdotes, a fable about the bond of brotherhood:

> If a philosophical community still pictures itself as an all-male compact, and imagines itself to have been so since the early dawn of our subject, then it can go on producing crazy tales like that one, as its emblems, so to speak. And at the end of the day, we would have only biased "theories" about many political concepts which are important in society at large, and no critical means to challenge them properly.[84]

Questioning which emblems are remembered, naturalized, deemed inevitable—this is one way that feminist criticism has broached the historical chasm of women's inequality. Must the voice of language science always sound forth from a male mouth? Must daughters be subordinate to their fathers? Must one reactivate the same stories, again and again, to explain what language is and how we speak it? These are questions that the authors examined in this chapter implicitly pose as they negotiate the gendered tropes that had, by the mid-nineteenth century, become de rigueur for conceptualizing languages and their speakers. While some, such as Carolina Michaëlis de Vasconcellos and Elisa Richter, challenge the gendered metaphors and iconography that cast the human subject as specifically male, others, like Clementine von Münchhausen and Carla Wenckebach, perpetuate the gendered conventions that made possible their own exclusion. In this way, the history of women writing about language does not always overlap with the history of feminism. Whatever path they chose, however, in both their study and production of language, these thinkers had to reckon with the gendered conceptual sediment that the nineteenth-century discourse on language left in its wake.

Modernism's Masculine Language Crisis

That Germanophone literature and philosophy underwent a "language crisis" in the early twentieth century is now such a commonplace that it has become a point of scholarly complaint.[1] The "so-called language crisis," as it is so often named, is associated with figures such as Fritz Mauthner, Ludwig Wittgenstein, and Hugo von Hofmannsthal, writers whose works thematize a skepticism about language's ability to adequately capture and express experience. Whether we choose the term language *crisis* or language *skepticism* or something else altogether, however, is less important than recognizing that many texts of this period are undeniably preoccupied with language—its expressive limits and possibilities, the way it structures knowledge, and whether it can truly represent or communicate human experience. Such questions signal a reaction against the conceptualization of language put forth by Humboldt and his contemporaries, where language was conceived as an autonomous organism in which man recognizes the structures that organize his own life. Or as we saw Lia Formigari describe this perspective in chapter 1, although man may be a "prisoner of language," he is a happy prisoner, since language "is not foreign to the true original nature of man."[2] Around 1900, linguistics and language philosophy turn away from that conceptualization of language. This is expressed most famously in Ferdinand de Saussure's notion of the arbitrariness of the linguistic sign, but is also apparent in the Neogrammarian critique of Grimm encountered in chapter 2 (Brugmann was in fact Saussure's teacher).[3] For Saussure, as Sarah Pourciau writes, language "does not point outward toward things or inward toward thoughts [. . .] it means, but it does not do so in the direction of something one could call truth."[4] In 1922, Ludwig Wittgenstein would famously write that "the limits of my language mean the limits of my world."[5] Humboldt might have said something similar, but with a fundamentally different valence. For Humboldt, it is precisely these limits that make authentic self-expression possible (everything we are and might want to be is already contained within language). In Wittgenstein's *Tractatus*, on the other hand, the limits of language suggest that there are things that language cannot access: "What we cannot speak about we must pass over in silence."[6]

Literary texts in this period, too, reckon with end of the "Romantic" conception of language, which had posited a mutually affirmative relationship between language and the human. In doing so, they also reckon with the gendered framework upon which that conception was built. As I argued in the first three chapters of this book, in nineteenth-century linguistic thought, "the human" was a category from which women were excluded. In the early twentieth century, literary authors address and exploit this exclusion, articulating their "crisis" of language as a *masculine* crisis to which feminine language is posited as a solution. In the poetic meditation "Das Gespräch" ("The Conversation"), for example, Walter Benjamin casts "die Sprache der Frauen" (the language of women) as an antidote to regular language, which he describes as instrumentalizing and incapable of facilitating true communication. Rather than being an inferior version of men's language—as it was so often characterized in the previous century—in the texts of Benjamin and his contemporaries, "women's language" becomes a utopic alternative to the limitations that regular language imposes.

This chapter explores how and why the "language crisis" was figured as a masculine language crisis through a focus on three case studies: Hugo von Hofmannsthal, Robert Musil, and Walter Benjamin, particularly the work they produced in the first quarter of the twentieth century. In this, they are situated at the tail end of my book's period of focus. I concentrate on their work rather than on literature from the nineteenth century because they most directly disclose the borders of the masculinist conception of language outlined in this book. Hofmannsthal, Musil, and Benjamin do not simply duplicate preexisting notions of the alterity of women's speech but radicalize them, push them to their extreme, and probe their capacity to generate a new poetics. In their texts, "women's language" becomes something totally incomprehensible—a "magic language," a "language of the whole," a language of "pure" bodily expression. In rendering "women's language" this way, the authors both reproduce and critically reflect upon the sexual differentiation performed in earlier linguistic theory. Insofar as they entertain the possibility that "women's language" could be a solution to the problems of "male language," the texts practice a kind of benevolent misogyny, in which woman, although now valorized, can only be the artist's inspiration rather than the artist herself. Silence remains the feminine ideal, as it was characterized in earlier centuries. Yet while they perform this move, the authors—Musil and Benjamin and, to a lesser extent, Hofmannsthal—also interrogate it, suggesting that very *idea* of "women's language" is not natural but constructed, mediated by what Sigrid Weigel, writing on Benjamin's work of "decipherment," has called "the phantasmagorias, the wish-symbols, and the materialized images of the collective."[7]

Hofmannsthal's Female Bodies

Although he began his literary career as a poet, the Viennese author Hugo von Hofmannsthal also went on to compose pantomime and dance librettos, to write a theoretical text on pantomime, to collaborate with dancers of his time, and to incorporate scenes of dance into his dramas. Critics have documented how Hofmannsthal's interest in dance can be understood as a response to the problems of language described in his earlier work, a "medial shift from the word to the body" in Carsten Zelle's phrase, or "body language and crisis of language" in the words of Alys George.[8]

What I want to show in this section is that Hofmannsthal's turn from the word to the body is not just a turn to the body in general, but to the female body in particular. Or to put it more precisely: Hofmannsthal's turn to dance and the body is *facilitated* by a certain idea of "woman" that understands woman to be closer to nature, more tied to the body, than her male counterpart. Of course, many of Hofmannsthal's texts on dance include male as well as female characters, but in the places where the idea of a nonarbitrary sign is most explicitly being worked out, and where the dance is supposed to be completely expressive, nonrepresentational, Hofmannsthal takes recourse to female figures and tropes of femininity.

Hofmannsthal most clearly and famously articulated his language skepticism in the 1902 prose text "Ein Brief" ("A Letter"). This letter from the fictional Lord Chandos to Francis Bacon describes, in elegant prose, Chandos's increasing mistrust of language and his subsequent loss of speech. The performative contradiction that the letter enacts—How can a loss of language be rendered *in* language?—is never resolved within the text. Hofmannsthal has Lord Chandos conclude his missive by writing about a new language "in which inanimate things speak to me," translating his silence into intelligible language for the reader.[9]

Ten years later, in the theoretical text "Über die Pantomime" (On pantomime), Hofmannsthal resolves the tension between form and content staged in "A Letter" by turning from language to the body, and from narrative prose to dance and pantomime. Pantomime was a common interest of the era; the revival of this art form was, in George's observation, "a centerpiece of Viennese efforts to reconceive modernist literature at the tail end of the nineteenth century."[10] Taking inspiration from a dialogue by Lucian on the same subject,[11] Hofmannsthal reconfigures the ancient Greek text to address his specific concerns about the limits of linguistic expression and the corresponding need for nonlinguistic aesthetic media. In "Über die Pantomime," Hofmannsthal emphasizes dance's capacity to portray what language cannot. Like language, pantomime does include representational elements, Hofmannsthal argues, but it cannot be reduced to these elements. Pantomime would be unthinkable

> without its being permeated through and through by the rhythmic, the purely dance-like; failing that, we find ourselves in a play whose performers absurdly use their hands instead of their tongues; in other words, in an arbitrarily irrational world that is oppressive to remain in. On the other hand, condensing a state of mind into a stance, a rhythmic repetition of movements, therein a relation to surrounding people, more dense and more significant than language, would be able to express, to reveal something that is too grand, too general, too close to be put into words.[12]

This emphasis on pantomime's nonrepresentational nature is not in Hofmannsthal's source text. Lucian finds no fault with pantomime as imitation, only when this imitation exceeds the proper bounds and becomes too exaggerated.[13] And while Lucian does compare pantomime to other genres in order to defend its legitimacy, he makes no argument for dance overcoming the impotency of language. This is Hofmannsthal's own addition, reflective of his particularly modern concern with language as arbitrary and artificial.

If the world of words is limited by arbitrariness, as Hofmannsthal suggests, then the world of pantomime is supposed to transcend this arbitrariness because its medium is not language but the body. The rhythmic and "purely dance-like" gestures of pantomime are what allow pantomime to serve as a countermodel to language because they are not imitative but simply expressive. By championing dance here, Hofmannsthal is able to offer in "Über die Pantomime" something that was not possible in "Ein Brief"—an alternative to language other than silence.

This alternative medium of expression, of course, still exists beyond the bounds of his own text. If dance exceeds the linguistic, then it is impossible for Hofmannsthal to adequately describe it in prose. Indeed, in "Die unvergleichliche Tänzerin" (The incomparable dancer), an essay on the dancer Ruth St. Denis, Hofmannsthal designates her performance as "unportrayable" and "indescribable," resolving that he will "barely try to describe her dancing."[14]

Yet the difference between "Ein Brief" and these texts on dance is that "Ein Brief" posits a resolution to the problem of language *within* language (the creation of a new language in which "silent things" communicate, which cannot currently be manifested in the world but exists only in the realm of the theoretical), while "Über die Pantomime" points outside of itself to locate the solution in a different aesthetic medium.

Before investigating the work that "woman" performs for Hofmannsthal's conception of dance, it is important to clarify that Hofmannsthal is interested in a specific kind of dance—not formalized European dances, like the waltz, but the more experimental movements of modernist dancers like Isadora Duncan, Ruth St. Denis, and Grete Wiesenthal.[15] For Hofmannsthal, this kind of dance is associated with practices of ancient and foreign cultures. As he puts it in "Über die Pantomime,"

> This form of expression is common in simple heroic times, indeed especially in the primeval state. However, since everything that is human persists, from the multiplicity of our overarching present the same indestructible need arises, which art seeks to soothe by offering us one of its primeval forms for new animation, since the foundation of life is unfavorable.[16]

Similarly, what Hofmannsthal values in Ruth St. Denis's "oriental dance"[17] is that it is "something so thoroughly strange and that is in no way ashamed of its mysterious strangeness; that seeks no mediation, no bridge; that wants nothing to do with education, doesn't want to illustrate anything, doesn't want to make anything accessible."[18] The alterity that distinguishes St. Denis's dance also marks her body. Although Hofmannsthal writes that she is either Canadian or Australian (in fact she was American), he muses that she must have "another drop of more foreign blood, a grandmother of Indian blood, something with the mystery and power of a primeval race."[19] While part of a different theoretical and aesthetic program, this move—the conflation of linguistic, sexual, and racial difference—is reminiscent of the nineteenth-century construction of "women's language." There, the desubjectification of both "women" and racialized "primitives" was performed by asserting their interrelation. While Hofmannsthal valorizes rather than denigrates this "oriental dance," "Über die Pantomime" relies on a similar conception of otherness, in which alterity exists as a fungible commodity that can be exchanged among and represented by various marginal identities. Toni Morrison, writing on the function of Blackness in the American literary imagination, calls this "metaphysical condensation": a strategy of representation that "transform[s] social and historical differences into universal differences."[20] For white American authors, Morrison argues, "the slave population, it could be and was assumed, offered itself up as surrogate selves for meditation on problems of human freedom."[21] In Hofmannsthal's work, it is "woman" who is offered up as a surrogate for meditation on the problems of language.

In order to look more closely at the function of "woman" in Hofmannsthal's notion of dance, I turn now to his 1907 work *Furcht* (*Fear*). This is a fictional dialogue between two female courtesans that presents, perhaps more extensively than any of his other texts, his conception of dance as essentially nonrepresentational. Like "Über die Pantomime," *Fear* also takes inspiration from Lucian—here it is his dialogues of the Hetarae, the ancient Greek courtesans, upon which Hofmannsthal draws.[22] Yet, as he also will do in "Über die Pantomime" a few years later, Hofmannsthal deviates significantly from Lucian's text, using the ancient dialogues only as a springboard from which to explore his own concerns—namely, the expressive capabilities of different aesthetic media.

In *Fear*, Hofmannsthal stages a discussion between two courtesans who disagree about the virtues of the dance they perform for their male audience. While the first dancer, Hymnis, is relatively uncritical of their manner of dancing, the second dancer, Laidion, is dissatisfied, arguing that the performance is only for the benefit of the men and never makes her happy. "My God," Laidion proclaims,

> How flat and futile all that is! There we dance for twelve or twenty men, among them a few rich old dodderers and the rest parasites. We dance and then we are tired and then everything turns ugly: everything crowds in on me, the faces of the men, the lights, the noise, like the beaks of greedy birds everything pecks me in the face. I would rather die than lie with them and drink and listen to their shouting. Then I wish myself as far away as a bird can fly.[23]

In describing her disgust for their current manner of dancing, Laidion makes use of two bird comparisons: the men peck at her face "like greedy bird beaks," and she wants to go as far away "as a bird can fly." The way the simile functions here—that the men are *like*, but not actually the *same as* birds—is precisely what Laidion criticizes about the dance that she and Hymnis perform. "But you have wishes," she tells Hymnis, "and wishes are fear."

> All your dancing is nothing but wishing and striving. You spring to and fro: are you fleeing from yourself? When you are hiding, are you hiding from the eternal restless yearning within you? You are aping the gestures of animals and trees: do you ever become one with them? You step out of your garment: do you step out of your fear? Can you ever, even for two hours, lose all fear?[24]

The problem with the way the women dance, according to Laidion, is that their dance is mimetic. The dancer may ape the movements of animals and

trees, but she will never be an animal or a tree; this difference between signifier and signified, between the self and the world, and between the dancer
and the dance she performs, is never eliminated. Why is this a problem? For
one, Laidion's argument implies, because it is restrictive: the dancer is ruled
by the desire of the audience, rather than by a desire to give herself over to
the dance itself. Elsewhere, Laidion compares herself and the other dancers
to marionettes. The male spectator "holds the strings up there, fastened to
the centres of our bodies, and pull us to and fro and makes our limbs fly."[25] It
is not desire, Laidion continues, that makes her perform the ecstatic movements of a maenad, but fear. In other words, their dance does not allow for
authentic expression. They dance for the sake of the men, rather than for the
sake of the dance.

Even more importantly, the critique that Hofmannsthal ventriloquizes
through Laidion is that the type of dance they perform does not allow dance
to express its true form. Instead of dance in the emphatic sense, we have here
dance masquerading as language—dance that attempts to signify and convey
meaning, as language does. Dance, however, has an alternative available to it
that language does not: the movements of the body, gestures which do not
stand for anything external but are "pure" expression.

Hofmannsthal uses this alternative conception of dance to construct a
countermodel to Hymnis and Laidion's performance. This countermodel,
this "dream of another dance," to use Gabriele Brandstetter's phrase, is a
dance that Laidion has heard is performed on a distant island.[26] According
to a story told by a visiting sailor, there is an island where "primitive" people
live and worship their gods through dance. These people have an animistic
culture that makes no distinction between religion and nature. The trees are
giant and their shadows are "like something alive," while the gods live "in the
trees and between the trees."[27] The people there have no shame in their dancing, Laidion explains. The men crouch on the ground and the girls "stand in
front of them all together, so motionless that their bodies are like one body.
Then they dance and in the end they give themselves to the youths, without
choosing—whoever seizes one, his she is. For the sake of the gods they do it
and the gods bless it."[28]

While both kinds of dance lead to sex (the dance Laidion and Hymnis
perform concludes by "laying with" the male spectators), Laidion's description of the island dance turns sex into a religious fertility ritual rather than
crude secular amusement. Furthermore, both the dance and the sex that this
dance sanctifies are marked by nonindividuation. Instead of being individual
agents, the island dancers' bodies are "like one body"; when they go to reproduce with the men, it is "without choosing." The principle of differentiation,

which language depends upon, and which Laidion finds so oppressive in the courtesan dance she and Hymnis perform—you mimic a tree but never become a tree, for instance—this differentiation evaporates in the dance of the island. Laidion continues the description:

> Like you there the men sit, very small, very far away. And in the trees hang the gods, fearful, with wide-open eyes, but into those who have risen from the clean mats nothing can instill fear. They are charmed. Everything—fear, desire, all choice, all unquenchable restlessness—everything has been transformed at the limits of their bodies. They are virgins and have forgotten it, they are to become women and mothers and have forgotten it: to them everything is ineffable. And then they dance.[29]

While the men apparently do watch the women's dance on the island, Laidion's description makes it clear that the dance is not directed at the male spectators. The theme of lack of individuation is further developed here: the women forget that they will be wives and mothers, and instead exist in a kind of urfemininity before the establishment and institutionalization of social gender roles. As Brandstetter points out, the way that the dancer's bodies are supposed to blend together corresponds to a popular contemporary thesis about "primitive" cultures, which held that "in comparison to the civilized, the savage has a different rhythm of life, based on a different sense of self."[30] In the description of this dance, Hofmannsthal relies on two forms of difference—that of "primitive" cultures and that of the female sex—to signify the alterity of the dance itself.

Hofmannsthal ends *Fear* by having Laidion perform this other dance, and here the text breaks out of the dialogic form into something resembling stage directions. An italicized narrative voice, absent until this moment, enters:

> *She begins to sway from the hips. Somehow one feels that she is not alone, that many of her kind are around her and that all are dancing at once under the eyes of their gods. They dance and circle as dusk falls [. . .] They are givers of birth and the newly born of the island, they are the bearers [Trägerinnen] of death and life.*
>
> *At this moment Laidion hardly resembles herself any longer. Under her tense features appears something terrifying, threatening, eternal: the face of a barbarian deity. Her arms rise up and down in a frightening rhythm, death-threatening, like clubs. And her eyes seem filled with a hardly bearable tension of bliss.[31]*

In this ecstatic moment of performance, Laidion is transported out of herself and onto the island; she dances as one with other women. This dance is marked as specifically feminine. Its performers are described in the feminine plural, the "Trägerinnen des Todes und des Lebens" (bearers of death and life).

Hofmannsthal would thus appear to find in the expressive and "frightening rhythm" of the women's dance a tidy solution to the problems of linguistic representation that he so often thematized, except for the fact that *Furcht* is not a performance piece. It is one of Hofmannsthal's many *erfundene Gespräche*, fictional conversations or dialogues. Insofar as the text gives us a description of dance rather than dance itself, it marks the notion of the dance's alterity as a kind of conceit, an idea *invented* by the text rather than something the text can perform. The reader does not participate in Laidion's ecstatic scene but remains a passive spectator to an imagined event. In this, the text seems to recognize the "dream of another dance" as precisely that—an unrealizable fantasy.

Nevertheless, Hofmannsthal did write performance pieces where he staged this kind of dance. In the tragic drama, and then opera, *Elektra* (*Electra*; 1903/1909), for instance, Hofmannsthal has his female protagonist perform movements similar to that of Laidion. Basing his *Elektra* on the Sophoclean tragedy, Hofmannsthal maintains the general storyline and characters from the ancient Greek. As many scholars have shown, he also incorporates elements of the early twentieth-century discourse on hysteria, led by Freud and Breuer's studies, into his portrayal of the tragedy's female characters.[32] I am less interested here in the role of hysteria, which has been well documented, than on the way that Elektra's dance is configured in contradistinction to the linguistic. Both the stage directions and the way that Elektra speaks of her dance emphasize its extralinguistic nature. "*Electra has risen*," Hofmannsthal's stage directions read, "*She comes striding down from the doorsill. She has thrown back her head like a maenad. She flings her knees up high, she stretches her arms out wide, it is a nameless dance in which she strides forward.*"[33] The movements of Elektra's body echo those of Laidion, who also throws her head back and moves her arms in the same manner. As Elektra performs this "nameless" dance, she proclaims,

> All must
> approach! Here join behind me! I bear [*trag*] the burden
> of happiness, and I dance before you.
> For him who is happy as we, it behooves him to do
> only this: to be silent and dance![34]

In both *Fear* and *Elektra*, Hofmannsthal correlates this ecstatic type of dance with the word *Tragen*. Elektra declares that she bears, or *trägt*, "the burden of happiness," while the description of Laidion's dance refers to the women as the "Trägerinnen des Todes und des Lebens." For Hofmannsthal, the body of the dancer does not represent, signify, or even *perform* (for that would be too spectator-oriented, like the repudiated dance of the courtesans in *Fear*), but *trägt*, carries the experience that language fails to express and that can only be externalized through movement. A second meaning of *tragen* is operative here, that of biological reproduction: *tragen* also means to bear or yield fruit, as well as (for female animals) to be pregnant. In this way, it is not a coincidence that Hofmannsthal chooses to imagine a female fertility ritual in *Fear*, his most programmatic text on dance. The dancers there are the "birthers and newly born" not only because they are about to become mothers and have children but also because the dance itself is productive, bringing forth new expression. That he continually makes use of female characters and the idea of Tragen in his depiction of this "other" dance suggests that, for Hofmannsthal, the medium of dance is inextricably linked to the maternity, to what Brandstetter has called a "female creativity."[35] Indeed, we can see the roots of this idealization of the maternal already in the Chandos letter. As Hofmannsthal describes the lost language that Chandos desires to recuperate, he engages in what David Wellbery terms a "lactopoetics," using imagery of the infant's connection the maternal breast to express the fantasy of a language that would speak the subject undivided from the world around him. This linguistic fantasy is, Wellbery suggests, *incestuous*, and we can understand Chandos's so-called "language crisis" as its punishment—an oral punishment for an oral transgression.[36] While Hofmannsthal does not continue this particular emphasis on the maternal breast in his later writings on dance, he does continue to take recourse to the female body as a means of representing problems of language.

The association of female characters with nonrepresentational movements, which define the dances of *Fear* and *Elektra*, also appears in Hofmannsthal's pantomimes. In *Das fremde Mädchen* (The foreign girl), from 1910, the most indecipherable gestures of the piece are assigned to the female character from whom the pantomime takes its name.[37] In *Amor und Psyche* (Amor and Psyche), written the following year, Hofmannsthal reworks the classical love story and devotes the majority of the pantomime to the character Psyche. In her most expressive dance, Psyche's movements parallel those of Laidion and Elektra. "Is outside and inside the same?" Hofmannsthal's description reads, "the poor head can no longer tell the one from the other: a fear-ridden desire to get out of this circle of hell, a falling forward—then she doubles up, throws herself upwards, as if she wanted to throw herself out of herself."[38]

To Grete Wiesenthal, the dancer who would perform the lead role in both *Das fremde Mädchen* and *Amor und Psyche*, Hofmannsthal wrote about his directions for the pantomime: "These are words. You translate it for yourself into a better material."[39] What exactly makes the body a "better material" than language? In *Fear*, Hofmannsthal critiqued the courtesans' dance for not being true to its own form, for trying to mimic language and represent rather than simply express. This distinction again appears in Hofmannsthal's letter to Wiesenthal, where he distinguishes between "pure" and "impure" gestures:

> It has to do with a series of *pure* postures and gestures. The gestures that accompany acting are all impure because they blend into each other; they merge with each other; they are also impure by their nature, to a small extent they are truly developed mimic gestures, for the most part conventional signs, like letters, which have developed from true images, hieroglyphs.[40]

Unlike words, which are arbitrary, tethered to their meanings only by convention, the body for Hofmannsthal offers the possibility of a different kind of sign: a sign that is motivated, whose form is ostensibly necessitated by nature, and which Hofmannsthal therefore conceptualizes as "pure." As Hofmannsthal writes in a different letter to Wiesenthal,

> Gretl, the present is nowhere and everywhere, the secrets are apparent, the deeds obscure, but pure because they want only themselves and are enclosed in themselves. Words confuse and go over from one thing to another; they are dangerous because they are without a self and wander out of themselves.[41]

Language can be deceptive, Hofmannsthal implies, because it has no *necessary* form; it is "without a self." The gestural movements of dance, on the other hand, precisely because they are unintelligible, ostensibly exist only for themselves. The instances of ecstatic dance from *Fear* to *Amor und Psyche* thus offer the body as an alternative to what Hofmannsthal would elsewhere call the "deceptive" nature of language. Unlike words, the frenzied movements of the dance do not pretend to *mean*, a disingenuous standing in for something else, but are "pure" expression.[42] What makes it possible for Hofmannsthal to gender this other dance feminine is the presumption that woman, more than man, bears a connection to the primitive culture out of which this dance supposedly originates. The comparability of "the woman" and "the primitive" was a well-trodden path in the previous century, asserted by Darwin and his contemporary scientists, as we saw in chapter 2. The idea that woman always retains a link to a lower stage of culture was also popularized by Johann Jakob Bachofen's mid-nineteenth-century monograph *Das Mutterrecht* (*Mother Right*). This text, which Hofmannsthal knew, argues that the order of modern civilization only came into effect once men overthrew the rule of matriarchy to establish paternity as law.[43]

In Bachofen, *mother* and *father right* refer not only to historical epochs but also to their attendant epistemologies, which are based in the different relational models of maternity versus paternity. Whereas the mother who births the child is (at least in an age before surrogacy) "always a physical certainty," the father is a "remoter potency" whose relationship to the child can never be known for sure. Grounded as maternity is in what can physically be observed, the matriarchal understanding of the world is "entirely subservient to matter and to the phenomena of natural life, from which it derives the laws of its inner and outward existence."[44] Patriarchy, on the other hand, "brings with it the liberation of the spirit from the manifestations of nature, a sublimation of human existence over the laws of material life." "Triumphant paternity," Bachofen writes, "partakes of the heavenly light, while childbearing mother-hood is bound up with the earth that bears all things."[45] "Father right" makes this heavenly light possible because it introduces abstraction—paternity is a juridical *idea* rather than a fact of nature—and this paves the way for the principles of abstraction and substitution that make higher civilization pos-sible. Hofmannsthal does not produce a systematic theory of history like *Das Mutterrecht*, nor does he praise patriarchy in such rhapsodic terms. One place these texts converge, though, is in the association of femininity—and maternal femininity, in particular—with a kind of natural semiotics.[46] For Bachofen, this is the shortcoming of the maternal order: under mother right, meaning is determined by (and bound to) to the observable rhythms of the natural world. In Hofmannsthal, on the other hand, this boundedness pro-vides a longed-for certainty. Unlike the arbitrary signs of regular, "masculine" language, which have no guarantor for their forms except convention, the forms of "feminine" language are motivated by the structure of the body.

Yet Hofmannsthal's texts also appear unsure about what such an under-standing of "feminine language" can really achieve. In *Fear*, as we saw, the liberatory capacity ascribed to the feminine "other dance" is acknowledged as constructed insofar as the text does not allow it to be performed, but relegates it to a linguistic *representation* of performance. Even *Elektra*, which stages such a performance, places clear limits on how this performance works. The drama ends not with Elektra's dance but with a line from her sister, Chryso-themis, who knocks on the door of the house, yelling the name of their brother. Chrysothemis, a character who happily accepts a fate of marriage and childbirth, had earlier declared her aversion to Elektra's plan of revenge by explaining, "I am a woman and I desire a woman's fate."[47] As Chrysothemis yells out "Orestes," the drama's conclusion coincides with a return to mas-culine authority. Elektra lies collapsed on the stage; her ecstatic dance, the activity of tragen, can only be realized momentarily before it is overcome by a more conventional order of representation. "Feminine" dance may be posited as a solution to the limitations of "masculine" language, but it is also marked as a transitory illusion.

Feminine Language and Narrative Form in Musil

In his early work, Musil crafts feminine figures who have a strange, alternative relationship to the world—and, correspondingly, an alternative medium of expression. In the 1911 collection *Vereinigungen* (*Unions*), female characters are said to have access to language that is different from the regular language of the everyday: in the first novella, "The Culmination of Love," this is a language of the body, while in the second, "The Temptation of Silent Veronica," the protagonist is described as speaking in the language of angels. In the 1924 novellas "Grigia" and "Tonka" from the collection *Three Women*, the eponymous female characters are said to speak "magic words" and a "language of totality," respectively.[48] In what follows, I explore not only how these texts adapt preexisting ideas about "women's language" but also the critique that their adaptation poses to the formulations of language and sexual differentiation operative in nineteenth-century language science and philosophy. In particular, Musil shows the very *idea* of a separate "women's language" to be a masculinist fantasy.

Like Hofmannsthal, Musil was interested in the relationship between language and experience, in whether language has the capacity to represent all aspects of life. His first novel, the 1906 *The Confusions of Young Törless*, takes as its epilogue a quotation from the writer Maurice Maeterlinck, which declares, "In some strange way, we devalue things as soon as we give utterance to them."[49] Maeterlinck expresses here "the obscure presentment of a realm before language and thought that is presumed to provide the ultimate foundation for ordinary experience even as it eludes conceptualization," to use Patrizia McBride's summary.[50] McBride and other critics have suggested that Musil does not depict a "language crisis" per se, but rather an exploration of "an ineffable ethical realm": "Ordinary language and thought may not do justice to the preconscious life of the soul, but they still provide a valid means for coming to terms with its elusiveness."[51] In other words, in Musil, the problem is not the inability of language to represent at all, but rather a particular kind of mystical experience in which the ordinary structures of the world no longer operate.

In 1925, Musil would describe this "preconscious life of the soul" as "the Other Condition" (*der andere Zustand*). Writing in the essay "Toward a New Aesthetic," Musil characterizes this as a fleeting yet fundamental human condition in which

> the presence of another world, like a solid ocean bottom from which the restless waves of the ordinary world have drawn back; and in the image of this world there is neither measure nor precision, neither purpose nor cause: good

and evil simply fall away, without any pretense of superiority, and in place of all these relations enters a secret rising and ebbing of our being with that of things and other people.[52]

The "Other Condition" poses a problem to representational literature, indeed to linguistic representation generally, because it does not recognize differentiation—a principle upon which language of course depends. Later, Musil will focus his energies on this problematic in his magnum opus, *The Man without Qualities*. In her work on Modernism's "primitive thinking," Nicola Gess argues that this novel uses figures of animals and madness in order to explore "a defamiliarized version of the familiar and an alternative relationship to the world."[53] As the following section details, however, in Musil's early novellas, the figure of choice for this exploration is woman.

TEXTUAL UNIONS IN "THE CULMINATION OF A LOVE"

"The error of this book is that it is a book," Musil once wrote about his novella collection *Unions*. "That it has a cover, spine, pagination. You should spread out a few pages and change them from time to time. Then you would see what it is."[54] Musil wants to free his narrative from the constraints of narrative—the conventions of what he would later call the "old narrative naïveté"—which, he claims, are too restrictive to adequately express new ideas. One of these constraints is chronology. Instead of pagination, which implies a sequential ordering, Musil writes that the book's pages should be changed "from time to time," without regard for temporal progression. In this way, the narrative would contradict the very foundations of the genre. The book as it is imagined here is literature posing as visual art. This new way of writing would create a text that should be looked at rather than read, encountered rather than understood. Musil's notebook entries from around the time of the publication of *Unions* further emphasize his desire to break with realist modes of narration: "One could try to determine," he writes in one entry, "that these narratives are shaped by a disgust with narrative."[55] Elsewhere he describes *Unions* as a "concentration of almost mathematical rigor, tightest mosaic of thoughts,"[56] further encouraging the notion that his text should be viewed as visual art (in an instant, as mosaic) rather than as a narrative depiction of events in time.

On the face of it, Musil's radical descriptions of his work appear disingenuous. For he does not give up all the trappings of narrative prose in "The Culmination of a Love," and the events of the narrative do follow a general progression. The main character, Claudine, leaves her husband at home in the city and takes a train to visit her daughter, who is at school in the country.

Away on this trip, Claudine meets a man, the "undersecretary," for whom she feels simultaneous disgust and desire and to whom she is sexually drawn. The story ends as they consummate their relationship.

While it is possible to sketch an outline of the plot, the events of this plot are not the focus of the narrative. Repeatedly, once a particular setting is introduced, the narrative moves away from any grounding in time and space to instead describe, for pages at a time, a transformation that Claudine undergoes. Progressively, Claudine experiences the erasure of various forms of difference—the distinction between herself and the world, the divide of past from present, and the differentiation between words and sounds that makes language possible. For much of the text, Claudine exists in a kind of mystical fugue state, disassociated from the world around her. As the narrative stages this transformation, it mimics Claudine's loss of differentiation formally by collapsing binaries on different levels of the text. This plays out in (1) the production of similes and nominalized verbs, (2) the merging of the voice of the narrator with the perspective of Claudine, (3) the fusion of multiple temporalities, and (4) the unification, on the level of the plot, of Claudine with her husband via her relationship with the undersecretary.

None of these textual "unions," however, makes a claim to the absolute elimination of difference. In the simile, for instance, two disparate things are combined, such as "a flush of emotion; a wondrously sweet burst of bitterness, like a gust of wind that lifts from the sea"[57]—here a joining of the human psyche and the natural world. Yet the "like," the very mechanism that brings together the two sides of the simile, also stands between them, simultaneously maintaining their dissimilarity. This "both/and" procedure of the simile, in which difference is *both* erased *and* preserved, can be taken as a model for Musil's text and its experimentation with narrative form.

Musil's engagement with the issue of the binary through these textual unions is significant not only as it relates to the experience that Claudine supposedly undergoes, but also as it relates to narrative as such. There is a binary structure inherent to the genre of narrative: the distinction between narrator and narrated, which all third-person narration presupposes in its form, even as it obscures this distinction through techniques such as free indirect discourse. To the extent that narrative is the verbal presentation of an unfolding of events in time, then the structure of narrative is also premised upon a binary, that of beginning versus ending, two poles separated by the progression of linear chronology.[58] Musil's play with binarism on all levels of the text—in figurative language, perspective, plot—can therefore be understood as a recognition of and challenge to the underlying structure of narrative, and thus as a metatextual reflection on the possibilities and limits of the genre. This is the conceit of *Unions*, that it styles itself as an antinarrative but also retains a shell of narrative form, refusing to let go of the genre in its entirety, for this would lead to silence.

Having briefly sketched out the link between binarism and narrative form, we now come to the issue of sexual difference. The protagonist of "The Culmination of a Love" is a woman, I would like to suggest, not only because women and women's language have traditionally been associated with materiality and the body, two things that the text presents as constitutive of Claudine's experience, but also because sex is presumed to be one of the ultimate binaries of human life. In the bringing together of Claudine's point of view with what Musil stages as the "masculine" narrative voice, the text thus becomes a place where the two sexes are joined, and the supposed difference between them is elided. Yet insofar as it is Claudine's perspective, and not her speech, that is united with the narrative voice, Musil marks Claudine's relationship to language as distinct from that of the narrator and thereby maintains a differentiation between them. The masculine is articulate while the feminine is inarticulate; the masculine speaks while the feminine sees and experiences. To the extent the feminine is aligned with the visual here, it is also opposed to the "narratival," at least in Musil's constellation. If we remember from earlier, visuality was precisely the category that Musil used to imagine a new kind of narrative, a book that wouldn't be a book, a text that would approximate visual art. In constructing a feminine perspective that cannot be spoken, Musil's text attempts to stage this visuality, this antinarrative element, within itself. But the text also shows the impossibility of this project, as the narrator must constantly intercede, translating what is "seen" for the reader. The problem of what we might call "feminine language" thus appears in *Unions* less as a problem of sex or gender, or even as a problem of language, but as a problem of narration: How can a narrative include that which stands in opposition to the very premises of its own form?

To understand how Musil uses gender as a strategy to incorporate this "other" kind of language and experience into his text, consider the language of the text itself, specifically its use of similes and nominalized verbs. In an attempt to describe what is supposedly happening to Claudine, the text generates an extraordinary number of similes—the comparative structures *wie* (like), *wie wenn* (like when), and *als ob* (as if) appear 337 times, to be exact.[59] For example, in the following passage:

> Claudine's fidelity to her husband offered resistance, precisely because she did not feel it as a control but rather as a liberating force, a reciprocal support, an equilibrium achieved by the constant forward motion. A running hand

in hand, but sometimes she was gripped by the sudden temptation to stop midway, just her alone, to stop and look around. It was then that she felt her passion as something compulsive, coercive, overwhelming; and no sooner did she manage to subdue that cloying feeling than she was overcome with remorse and yet again infused with the consciousness of the beauty of her love, the yearning still lingered stiff and heavy like a frenzy, and she rapturously and fearfully sensed every movement she firmly entwined in its potent grip as if in a mesh of gold brocade; but something kept beckoning from somewhere, it lay still and pale as March shadows on the bare, broken ground of spring.[60]

The many comparisons seem to bring us closer to the experience presented, in that they provide another level of description and, therefore, insight. Yet insofar as the images produced take us out of the narrative setting—away *to* the shadow of a March sun, for instance—they also act as a distancing mechanism, reminding the reader that what the text is portraying here cannot be accessed directly, but can only be approached through these rather cliché comparisons. Claudine may be entering into a realm of experience in which difference no longer exists, but the text, in order to remain intelligible, must remain in the realm of differentiation. In its excessive generation of similes, then, the text asymptotically approaches what it presents as Claudine and her experience, but also marks its own limits, its inability to reach it entirely.

The presentation of time further underscores Musil's attempt to parallel Claudine's experience in the form of his text. Three different temporal modes are brought together in the passage quoted above: a turning backward to the past, a progressing forward into the future, and the sudden immediacy of the present. Elsewhere, the transformation that Claudine undergoes is presented as a return to an earlier state—she had the same experience once in America, when her daughter was conceived—suggesting that, for her character, the past is not superseded by the present but recurs in it. In the passage above, the three temporal modes (past, present, future) are also presented as coexisting for her at once, together, rather than being organized linearly. In its form, the text mimics this conflation of temporalities. Musil's narrative maintains the outline of a plot (in which events occur, time moves forward, there is a problem and a resolution), but also frequently disregards this plot by engaging in lengthy descriptions for pages at a time, which take the reader out of narrative time. That is to say, the structure of the text both imitates what it portrays to be Claudine's experience of time and, because it will not abandon a linear plot altogether, marks her experience as incompatible with narrative convention and thus impossible to completely incorporate into the text.

Like the simile, the compound nominalized verbs that Musil uses in the text perform both an erasure and an affirmation of difference. Nominalized verbs appear extensively throughout "The Culmination of a Love." To take just one passage as an example: Musil writes that after meeting the undersecretary, Claudine felt the encounter

> stirring something up in her, like when you walk by the seashore, unable to fully fathom the roar [*ein Sichuneindrückbarfühlen in dieses Tosen*] of every action and every thought torn in the fabric of the moment, and little by little she was gripped by a mounting uncertainty [*ein Unsicherwerden*] and a slowly growing inability to denote and sense the boundaries of self [*ein langsames Sich-nicht-mehr-begrenzen-können und-spüren*], a self dissolution [*ein Selbstvergließen*]—an urge to cry out, a longing for immeasurable movements, the rootless desire to do something unending, just to force yourself to feel; there was a sucking, lip smacking, devastating delight in getting lost [*Verlorengehn*], every second pulsing like a wild, irresponsible lonesome lust, devoid of memory, foolish and free. And it wrenched words and gestures from somewhere inside that flew by and yet remained a part of her, and seated there beside her, listening, the undersecretary was forced to fathom that what she said and did, all the words coming at him bore hidden traces of her beloved, and soon she saw nothing but the neverending rise and fall of his beard, the bobbing beard of a repulsive billy goat ceaselessly chewing, spitting out a whispered soporific stream of words.[61]

The German language allows for verbs to be turned into nouns with a simple capitalization; Musil pushes this possibility to the extreme here, crafting the words *Sichuneindrückbarfühlen*, *Unsicherwerden*, *Selbstverfließen*, and *Verlorengehn*. Writing these words as nouns brings together components that would otherwise be separated in the verbal form. By erasing the differentiation between words in this way, the language of the narrative appears to mimic Claudine's experience of disintegrating boundaries. The transformation of verb into noun, moreover, mirrors what the text presents as Claudine's nonlinear experience of time. Instead of an action (verb), a forward progression in time, we have a *thing* (noun), which stands outside of time, bearing no intrinsic relationship to any temporal mode. Yet Musil's creation of the hyphenated *Sich-nicht-mehr-begrenzen-können* (inability to denote the boundaries of self) also calls into question the ability of the narrative to unite itself completely with the experience it accords to Claudine. The form of this noun imitates what it signifies, in that the words themselves are no longer fully denoted, separated by spaces. Because the words are still individually identifiable, separated by hyphens, the form also underlines the gap between Claudine's experience and the narrative representation of it. The narrative may push the boundaries of language, constructing new words and generating endless figures of comparison, but it cannot actually narrate in the type of language that it has ascribed to Claudine, for to write in a way completely *unbegrenzt* would be to write without words or grammar, which would render the text incomprehensible.

In contrast to the language of the text, which is excessively figurative, and which maintains the principle of differentiation that makes meaning possible, the text presents Claudine as engaged with a language that is literal, material, and uninterested in signification. The "language" that the text ascribes to her character does not *mean*, but *is*. In the figure of Claudine, language is related to the body: in the quotation above, she wants to "cry out," and desires "immeasurable movements"—inarticulate, performative acts. Words and gestures are pulled from her like material objects. She listens to the talk of the undersecretary but understands his language only in terms of the bodily movement that produces it, the "neverending rise and fall of his beard." The notion that woman is more bodily than man, more material than intellect, is, of course, the age-old trope that facilitates this characterization.

Throughout the text, the experience of the world assigned to Claudine is continually marked as either prior to or beyond language. She is gripped by disgust, for instance, when she feels

> that it did not so much matter what people say of themselves, what they manage to put into words, but rather that any real justification was conveyed in altogether different ways—a smile, a lapse into silence, an ear turned inward to the secret murmurings of self. And she suddenly felt an inexpressible longing for the one other person, lonesome like her [. . .].[62]

Her "justification"—presumably of her adultery, although this is not made clear—is not to be found in words, but rather in a smile, in a falling silent, in a listening to oneself. Here, Claudine longs to express herself in something other than language (in the body, in silence); elsewhere, the text explicitly tells us that she experiences the world in a way that cannot be captured by words or concepts. As she is out on the street with the undersecretary, the narrator describes Claudine wondering:

> This life, all blue and dark and with a little yellow fleck . . . what does it really want? This cluck of chickens and quiet rustle of spilling feed through which life suddenly passes like a clock striking the hour . . . for whose ears does it sound? This wordless flux that eats its way into the distance and only from time to time, channeled through the narrow hiatus of a few seconds, flares up into a fleeting something, and otherwise lies fallow . . . what is it up to? She looked it all over with a silent gaze and sensed the things without thinking them, the way hands sometimes remain resting on a brow when there's nothing more to say.[63]

She "feels" the things around her but does not "think" them; she has lost the cat-
egorical structures that make language and thought possible. Indeed, Claudine
is almost always described as "feeling" rather than thinking, which would imply
concepts, order, form—all things that the text presents Claudine as having lost.

It is true that the text reports instances of Claudine speaking, but what she
says never completely makes sense. Her speech is often fragmentary, as she begins sentences that she cannot finish. At the very end of the text, for instance,
all she can do is say "I'm disgusted," expressing a corporeal reaction rather
than producing meaning.[64] (In the original German, "mir ekelt," she is furthermore the indirect object rather than the subject of the sentence.) Shortly
before this, when the undersecretary asks Claudine if she loves him, she can
only respond in a way that is adjacent to the question: "How strange that one
should like someone just because one likes him, his eyes, his tongue, not the
words, but the sound itself . . ."[65] What interests Claudine in language is not
the meaning it produces, but its sound and bodily origin, its physical source.

By the end of the narrative, this link between Claudine's language and her
body is made explicit. On the final page of the text, Musil writes,

> And Claudine kept quiet; only once more did she speak; while they were
> undressing; she started rambling, incongruously, perhaps senselessly, it was
> really nothing more than a painful caress over something [*Überetwashinst-*
> *reicheln*]: ". . . it's like walking along a narrow trail; animals, people, flowers,
> everything changes; you yourself become altogether different. You ask your-
> self: if I'd been living here from the start, what would I think about this, how
> would I feel it? How strange that all you have to do is cross a line. I'd like to
> kiss you and then leap right back and look; and then kiss you again. And every
> time I cross that line I'd have to feel it more distinctly. I'd grow ever more pale;
> the people would die around me, no, just shrivel up; and the trees and animals
> too. And all that there'd be left in the end would be a thin trail of smoke . . .
> and then nothing but a melody . . . passing through thin air . . . trailing over
> the emptiness . . ."[66]

It is not a coincidence that Claudine gives her final speech while undressing,
as this is the culmination of a process of merging language and body that
has been developing over the course of the narrative. The text likens Claudine's speech to a caress, an *Überetwashinstreicheln*—another nominalized
verb—so that Claudine's speech performs a movement that would ordinarily
be enacted by the hands.

The narrative, of course, cannot put an Überetwashinstreicheln into direct
discourse—it cannot say the nonsemantic or the bodily, constitutive parts of
Claudine's "other" language. In citing Claudine's speech here, the narrative
thus reduces her language to its own, and leaves out the very dimensions of

Claudine's language that would constitute its alterity. Instead of using wild gestures or screams, which the text ascribed to her earlier, Claudine speaks in similes here, and trades in clichés—we might say, she speaks as the narrator does. The attempt to capture this "other" language in direct discourse must necessarily fail, precisely because direct representation is not possible. Indeed, the narrative draws attention to this failure: in having Claudine's cited speech operate according to the principles of a naive realism, which assumes that this transgression, or the crossing of a boundary, can in fact be narrated, the text self-consciously exposes the problem of this approach. The attempt to narrate this "border crossing" is foreclosed upon as Claudine's reported speech ends in ellipses.[67]

In these ellipses, Musil's text gives way from language to punctuation, those marks that separate words and sentences and thereby clarify meaning. As Claudine's language retreats from the text, all that is left is the narratival ordering of her absent speech. Without words to organize, the form of narrative reveals itself here as pure structure. Thus a major principle of narrative form is laid bare: the sequential presentation of time. What is perhaps most radical about Musil's text, therefore, is not that it abandons the genre of narrative altogether—in the end, it *is* a book, not a painting or a mosaic—but that it presents the components of narrative for inspection, drawing our attention to the key elements of the genre.

"FEMININE LANGUAGE," IRONIZED

If *Unions* discloses the inner workings of narrative form, Musil's later collection *Three Women* reveals the premise that made this disclosure possible: the assumption that woman, with her ties to the natural world, has closer access to the experience of the ineffable than does man. Each novella of *Three Women* focuses on a male protagonist who, through free indirect discourse, is allied with the narrative voice and perspective. These misguided male characters imagine that the women they love speak strange, magical words, expressions of their strange, magical experiences. By ironizing the idea of an alternative feminine language in this way, Musil critiques the poetics of *Unions*, as well as the general idea of the alterity of women's speech.

The novella "Tonka" from *Three Women* describes an unnamed man who falls in love with a poor Czech-Austrian girl, "Tonka." The plot proceeds as follows: the man procures her a position as a caregiver for his grandmother, and after the grandmother dies, they move together to the city. Later, while the man is away, Tonka becomes pregnant and contracts a venereal disease. Tonka insists that she was never unfaithful, yet it remains a mystery how she could have become sick and pregnant in his absence if she did not sleep with another man. The protagonist finds it difficult to reconcile Tonka's vow of fidelity with the fact of her illness—"there was either some mystical bond linking him with Tonka or she was guilty on an ordinary human level"—and she eventually dies.[68] The division between scientific rationality and romantic idealism is clearly split along gendered lines in the narrative. The man is a student of chemistry who "turned a deaf ear to all questions that had no clear-cut answer,"[69] whereas Tonka is repeatedly referred to as existing in a "fairy-tale" rather than in reality.[70] As this brief summary makes clear, the narratives of *Three Women* do not have the same resistance to plot as those of *Unions*. They furthermore conform much more closely to generic convention, specifically the traditional definition of the novella as the depiction of an "unerhörte Begebenheit"—here that "unheard-of event" is the mystery of Tonka's illness and apparent immaculate conception.[71]

In addition to the notion that Tonka belongs to a fairy tale, the male protagonist idealizes Tonka in numerous other ways: he associates her with nature, he views her as simple and childlike, and he also believes her to speak a language that is different than his own. Of course, she does speak another language (Czech), but for the protagonist, the difference of her speech is not simply a matter of a foreign tongue but a language with an entirely different relation to the world. He concludes that she doesn't speak "the ordinary language" (*die gewöhnliche Sprache*) but rather "some language of the totality of things" (*irgend eine Sprache des Ganzen*).[72]

Something similar occurs in the novella "Grigia," which follows Homo, a geologist who leaves his family to travel to a village in northern Italy. There Homo takes part in the opening of a gold mine. Living in this village is a group of German speakers, one of whom, Grigia, becomes his lover. But both Grigia and Homo are married to other people, and Grigia's husband is, unsurprisingly, unhappy about her affair. The husband follows Grigia and Homo to a rendezvous in a cave and walls them in; Grigia manages to escape, but the text implies that Homo will die there. Homo associates the women of the village with everything that he, an engineer from a developed land, is not—the preindustrialized past and a more natural order of life.[73] Out of this sentimentalization of the village women comes Homo's affair, and he renames Grigia so as to make her better fit in with his fantasy. In reality, her name is "Lene Maria Lenzi," but he prefers to call her "Grigia, pronounced Greeja, after the cow she had, which she called Grigia, Grey One."[74]

In contrast to the way that Homo and the other men in the text use language—to assert their will upon the world[75]—Grigia's language is apparently magical:

And she talked a magical language [*Zauberworte*]. A nose she called a neb, and legs she called shanks. An apron was for her a napron. Once when he threatened not to come again, she laughed and said: "I'll bell thee!" And he did not know whether he was disconcerted or glad of it. She must have noticed that, for she asked: "Does it rue thee? Does it rue thee much?" Such words were like the patterns of the aprons and kerchiefs and the coloured border at the top of the stocking, already somewhat assimilated to the present because of having come so far, but still mysterious visitants. Her mouth was full of them, and when he kissed it he never knew whether he loved this woman or whether a miracle was being worked upon him and Grigia was only part of a mission linking him ever more closely with his beloved in eternity.[76]

Grigia does indeed speak "differently," in that she speaks a dialect rather than High German. But, as the narrative is aligned with Homo's voice and perspective here, it is Homo's sentimental imagination of Grigia that turns her words into "Zauberworte," her language into evidence of a miracle.

The notion of a different language, associated with the feminine, held positive potential in *Unions*: it was a way for the text to transcend the limits of its own language and narrative form. But here in *Three Women*, the idea of "feminine" language is no longer generative for the text's own poetics. Other related concepts from *Unions* also appear in a diminished capacity in *Three Women*, for instance the idea of ineffability. Musil uses the term *unsagbar* (unsayable) in "The Culmination of a Love" to refer to Claudine's ecstatic experience, which the text cannot capture. In "Grigia," *unsagbar* appears only to describe pieces of furniture. In Homo's room, "the beautiful mahogany beds were indescribably [*unsagbar*] cool and soft" and "the wallpaper had an indescribably [*unsagbar*] bewildering, maze-like pattern."[77] In the move from *Unions* to *Three Women*, the text's relation to the ineffable has, quite literally, been domesticated. The idea of *Unsagbarkeit* is flattened into banality—"unspeakable softness."

Whereas it was Claudine who was universalized "The Culmination of a Love"—by calling her an "everyday woman," Musil implied that she could stand for her sex in general—in "Grigia" and "Tonka," it's the male perspective that is constructed as universal. The male protagonist of "Tonka" has no name (he could be anyone) and the protagonist of "Grigia" is named "Homo," generic man. These narratives are less about the "three women" than about the male imagination of them. This shift in focus away from woman and toward the male gaze is also accompanied by a change in form. The narratives of *Three Women* are much less experimental than those of *Unions*. "Grigia" begins in an almost exaggeratedly conventional manner, with a maxim about life and then a description of Homo's family and setting.[78] "Tonka" begins in a way more reminiscent of *Unions*: "At a hedge. A bird was singing. And then the sun was somewhere down behind the bushes. The bird stopped singing. It was evening."[79] But this more radical narrative style is quickly foreclosed upon: "But had it really been like that at all? No, that was only what he had worked it up into later. That was the fairy-tale, and he could no longer tell the difference. In truth, of course, she had been living with her aunt at the time when he got to know her."[80] The narrative continues to correct itself in this way, providing a fairy-tale-like description and then revising it with the "truth," which corresponds to a linear narrative.

In contrast to *Unions*, the language of the female characters in *Three Women* is not a productive source for narrative invention. There is no tension between Grigia's supposed *Zauberworte* and the language of the narrative itself. Instead of attempting to mimic these Zauberworte, the narrative solves the problem of representation—How do you use regular language to depict magical language?—by shifting the burden of responsibility onto the male protagonist. The idea of Grigia's "other" language is not a premise that *the text* is positing (and thus would be invested in depicting), but only a fantasy held by one of its characters. And precisely because it has the status of fantasy in the text, the Zauberworte, just like Tonka's "Sprache des Ganzen," has no need to be manifested within the narrative.

"One does not express ideas in a novel or novella but rather lets them resound," Musil would write in a notebook entry shortly before the publication of *Unions*. "Why not choose the essay? Precisely because these ideas are not purely intellectual, but rather intellectual intertwined with emotional. Because it can be more powerful not to express such thoughts but to embody them."[81] The alterity of women's language is one idea that resounds in Musil's work. While *Unions* pushes conventional ideas about women's language to an extreme—rendering the alterity of *Weibersprache* not just a matter of grammar, vocabulary, or even intellectual content, but of the relationship between signifier and signified—"Tonka" deconstructs this notion, revealing it to be a figment of the male imagination. In doing so, Musil tries on, examines, and ultimately denaturalizes what was, in the previous century, an unquestionable premise: the intrinsic authority of the masculine position.

Uncomprehended Symbolism

In a 1913 text, the critic Walter Benjamin poses a curious set of questions: "How did Sappho and her female friends speak? How did women come to speak?"

> For language extinguishes their soul. Women receive no sounds from it and no salvation. Words waft over women, who are sitting together, but the wafting is

crude and toneless; they lapse into idle chatter. Yet their silence towers above their talk. Language does not bear women's souls aloft, because they do not confide in it [. . .] The words are mute. The language of women [*die Sprache der Frauen*] has remained uncreated. Talking women are possessed by a mad language.[82]

The notion that there are two fundamentally different kinds of language, one transcendent and one mundane, is ubiquitous in the Judeo-Christian tradition: we may think of the distinction between the language of God and the language of man, between human language before and after the Fall, or before and after the destruction of Babel. Indeed, these are distinctions that will interest Benjamin in later writings. But in this early text titled "Das Gespräch" ("The Conversation"), Benjamin makes no mention of God or of Adamic language. Instead, it is *women* who are associated with an alternative medium of expression.

"Alternative" is perhaps not a strong enough term. What Benjamin refers to here as the *Sprache der Frauen* is neither the ethnographically documented women's language of a "primitive" tribe, nor the cliché that women don't use language as well as men. It is instead a kind of antilanguage, immediate and nonconceptual, corporeal communion rather than semantic communication. Despite these discontinuities, however, the general structure of opposition into which Benjamin places women and "regular" language would have been at home in nineteenth-century language philosophy and science. In that framework, as we have seen, women's language could only ever be two things—an impoverished version of men's language (i.e., *Weibersprache*), or silence. But whereas nineteenth-century works repudiated Weibersprache as unrigorous, Benjamin proposes the "language of women" as an ideal with which his text gladly identifies and which it performatively attempts, through its own silences and ellipses, to approximate.

With its nonnarrative structure and opaque language, Benjamin's text is a challenge to summarize. The work is divided into eight parts; most sections consist of descriptive paragraphs, although there is also one dialogue. Parts 1 and 2 discuss exclusively male figures—fathers and youths, a male speaker (*der Sprechende*) and a male listener (*der Hörende*). As the text progresses, the figures increasingly move away from the male-dominated space of the first few pages. By section 4, the listener has been rendered female (*die Hörende*) while the speaker remains male. The conversation is now described in terms of heterosexual intercourse. The listener is likened to a prostitute who "receives" silence just as a woman "receives" a man. Then, at the end of section 6, Benjamin writes that men, whose language has been composed of dirty jokes and the violent manipulation of words, must "rise and smash their books and make off with a woman, since otherwise they will secretly strangle their souls."[83] Once the men stand up and close their books, they do not return—they are excised from the rest of the text. In the final two sections, men are replaced by "Sappho and her female friends," and heterosexual sex is replaced by the "unproductive" (*ohne Zeugung*) sex between women, as Benjamin turns to a contemplation of "the language of women." Just as Benjamin tells us that silence is the "internal frontier" of the conversation, women—and their language of silence—appears as the "internal frontier" of his own text, the point toward which his writing progresses.

According to "The Conversation," while women may speak in regular language (what Benjamin simply calls *die Sprache*), they cannot use it to truly express themselves and they also cannot control it. This is because women are not meant to speak it; or, alternatively, it is not meant for them. Woman's authentic language is something else, a language of silent bodily communion rather than verbal communication. Benjamin twice describes this language through the model of Sappho and her female friends, first in the passage cited at the beginning of this chapter, and then in the following description:

> Listen as they may, the words remain unspoken. They bring their bodies close and caress one another. Their conversation has freed itself from subject-matter and from language. Despite this it marks out a terrain. For only among them, and when they are together, does the conversation come to rest as part of the past. [. . .] Silent women are the speakers of what has been spoken. [. . .] None of them complain; they gaze in wonderment. The love of their bodies does not procreate, but their love is beautiful to see. And they venture to gaze at one another. It makes them catch their breath, while the words fall away in space.[84]

Benjamin directly contrasts "regular" language with the women's corporeal experience. While Sappho and her female friends either cannot or do not speak, they converse with their bodies, through "caressing one another." Benjamin works with a number of contradictions here. He writes repeatedly of the "language of women" and refers to the women's bodily communion as a "conversation," but their conversation is also clearly described as moving beyond anything that would be recognized as language. Furthermore, the women listen, but they are listening to silence; they speak, but they are speaking silence. Instead of words, which represent, the women have the body, which experiences. This experience is both haptic and optic (they "caress," "see," "gaze," "catch their breath"), but not symbolic, in that it does not stand for anything else. As Benjamin writes, their conversation has freed itself from both subject matter and language.

Benjamin's "Conversation" is not a conversation or dialogue in the traditional sense of the genre. The numbered sections do not represent speeches assigned to different participants, and there is no student and teacher or interlocutor and respondent. Instead, we might say that Benjamin stages a conversation between two kinds of language, what he portrays as the instrumentalized language of men and the silent language of women. Although the language of women cannot be manifested *in* regular language—and thus cannot be present in the content of the text—the text's form simulates its effect in its constant revisions and abstractions, preserving ambiguity and rendering full understanding impossible.

None of Benjamin's later, better-known texts on language—including "Über die Sprache überhaupt und über die Sprache des Menschen" ("On Language as Such and on the Language of Man"; 1916), "Über das mimetische Vermögen" ("On the Mimetic Faculty"; 1933), "Lehre vom Ähnlichen" ("Doctrine of the Similar"; 1933) or "Probleme in der Sprachsoziologie" ("Problems in the Sociology of Language"; 1935)—many of which have a hermetic style equal to that of "The Conversation," mention women or explicitly make use of concepts of femininity. This is not to say that Benjamin's thinking undergoes a complete transformation. In these texts there is still a distinction between what might crudely be called "good" versus "bad" language, or what Benjamin later distinguishes as the "bourgeois conception of language" versus "true language." In "On Language as Such and on the Language of Man," for instance, he makes a distinction between bourgeois instrumentalized language and Adamic naming language. Benjamin does write positively here of a silent language of materiality—but this is the silence of *objects*, not of women: "[Things] can communicate to one another only through a more or less material community."[85] Later, in the essay "Problems in the Sociology of Language," Benjamin also discusses the materiality of language in positive terms, but this is in reference to a "physiognomic phonetics," which focuses on the origins of speech within the body and believes that "articulation as the gesture of the speech organs falls within the large sphere of bodily mimicry."[86] Although Benjamin remains preoccupied with the relation between language, body, and materiality—and with "epistemological primitivism," to use Nicola Gess's term—he no longer expresses this relationship through women characters.[87] Benjamin's later work similarly maintains an interest in what Eva Geulen has called "the sexuality of language itself," or "an a priori linguistic sexuality": "A certain kind of reproductive or creative capacity [. . .] fertility, that language retained after God had removed himself from his creation, having released his all-creative word."[88] Yet this, too, is not discussed in conventionally gendered terms, or through figures of femininity.

It is impossible to say exactly why Benjamin abandons the practice of using female figures to explore the possibility of "another" language in his later
work, but we can see already in "The Conversation" an ambivalence about
the gendered approach that the text itself employs. In "The Conversation,"
Benjamin does not provide an explanation of why it is *woman* who holds the
status of the listener and speaker of a silent language. The assumption that
woman stands for everything that man is not, and that man commands the
dominant forms of expression, is something the text never bothers to justify.
And yet, as Sigrid Weigel points out, the text begins with the startling assertion that "what we do and think is filled with the being of our fathers and
ancestors. An uncomprehended symbolism unceremoniously enslaves us."
Although Weigel argues that the Benjamin of 1913 has not yet developed his
concept of dialectical images (that is, images that make their own "phantas-
magorical content" and "cultural constructedness" "readable"),[89] Benjamin is
clearly conscious—indeed, *self*-conscious—of the way that his text takes up
conventional, preestablished tropes, like the genius and the prostitute, and
thereby invokes the meanings that inhere within them.

We can see this in an early section of the text where Benjamin writes that
"silence is the inner frontier of conversation" and "the unproductive person
never gets to the frontier; he regards his conversations as monologues." This
same person, Benjamin continues, "saves himself by fleeing into the erotic. His
gaze deflowers. He wishes to see and hear himself, and for that reason he wishes
to gain control of those who see and hear."[90] These descriptions could equally
be self-accusations. Although titled a "conversation," whatever dialogue the text
presents is, of course, a fabricated one—a monologue of the author expressed
in a constructed form. Does not Benjamin, too, "flee into the erotic" when he
puts questions of language into sexual terms, when he solves the problems of
communication through the "caresses" of Sappho and her circle?

Benjamin was certainly aware of the limitations of using "woman" in this
way. In a personal letter written the same year as "The Conversation," he
explains:

> You should understand that I consider the types "man" and "woman" as some-
> what primitive in the thought of a civilized humanity. . . . Europe consists of
> individuals (in whom there are both masculine and feminine elements), not
> of men and women. . . . What do we really know of woman? As little as we do
> of youth. We have never yet experienced a female culture [*Kultur der Frau*],
> any more than we have ever known a youth culture. . . . To you, a prostitute
> is some kind of beautiful object. You respect her as you do the Mona Lisa. . . .
> But in so doing, you think nothing of depriving thousands of women of their
> souls and relegating them to an existence in an art gallery.[91]

Similar to how Benjamin describes the effect of language on Sappho and her friends, to behold women as works of art "deprives them of their souls." While "The Conversation" also employs such a strategy, the text acknowledges that it relies on a representation of women's difference that is not a given fact, but a construct.

"What we do and think is filled with the being of our fathers and ancestors," Benjamin proclaims at the beginning of "The Conversation." For someone like Herder, this was part of language's genius. That the sweat of the patriarchs is passed down through language is part of humanity's great patrimony. The writers examined in this chapter, on the other hand, are more ambivalent about the gendered figures of thought they inherit. They invoke these figures, but also recognize that they are not—as nineteenth-century linguistic thought might have it—natural and necessary, but culturally produced. The alterity of woman's language; the alterity of woman herself; the association of woman with the body, with sexuality, with nature: these tropes have always been phantasmagoric. But it is only in literature, that medium of phantasmagoric invention, that they are recognized as such.

Coda:
Gender and Language Today

"Hands off the generic masculine!" admonishes a 2018 article in the German newspaper *Der Tagesspiegel*.[1] Gender-equitable language reforms threaten "the highest social and political cultural heritage we have," declares a 2022 essay in the *Berliner Zeitung*.[2] These publications were penned by the linguist Peter Eisenberg, emeritus professor at the University of Potsdam. The linguist and Humboldt scholar Jürgen Trabant, retired from a professorship at the Free University of Berlin, made a similar intervention in *Die Welt* in 2021 when he criticized the so-called "political glottal stop."[3] This term refers to the practice of inserting a glottal stop within nouns in order to verbally express what would otherwise be noticeable only in writing: the "gender gap," which is used to render noun forms more inclusive. In German, nouns that refer to people are typically gendered based on the sex of those to whom they refer, with the masculine serving as the generic form. For example, *Studenten* (students) has historically been used for a group of students that included at least one man, while *Studentinnen* would refer to a group of exclusively women students. The *Binnen-I* (internal-I)—as in *StudentInnen*—was developed by feminists in the 1980s as an alternative to the generic masculine. Now, some speakers use forms such as *Student_innen* and *Student*innen*, where the "gap" provided by the underscore or asterisk functions to represent nonbinary people. In the introduction of these forms, and of the glottal stop used to pronounce them, Trabant sees nothing less than freedom under threat. It is "the most sophisticated piece of a fairly elaborate linguistic arsenal in the battle to conquer discursive superiority in the public sphere."[4]

It is telling that these essays reach their readers through newspapers rather than academic journals. Claims about "traditional" configurations of language and sex rarely inhabit today's scholarly works on language, though they live on in popular discourse. From concerns about the deleterious effect of women's vocal patterns like "uptalk" and "creaky voice,"[5] to scorn for gender-neutral pronouns, the presumed masculinity of standard language—or in the case of Trabant and Eisenberg, the presumption that masculine forms can include all genders without issue—has become a bulwark of conservative linguistic campaigns. Critics have railed against the introduction of gender-inclusive and gender-neutral words, arguing that they will damage the German language and, by extension, the German nation.

This criticism is not restricted to Germany, but is part of "the global right's antigender countermovement."[6] Gender-neutral language has become a contentious topic even in nations whose governments have encoded other LGBTQIA+ rights into law. In France, the introduction of the gender-neutral pronoun *iel* into a prominent dictionary in 2021 was met with charges of *wokisme* and calls to reject the importation of "American" "gender ideology."[7] That same year, the French education ministry banned the use of certain gender-neutral terms in schools.[8] Argentina, a country that "has largely embraced transgender rights" saw the city government in Buenos Aries "ban teachers from using any gender-neutral words during class and in communication with parents" in 2022.[9] In the United States and Britain, gender-inclusive language, like gender-neutral bathrooms and other material accommodations of trans and nonbinary people, has encountered a similar backlash.[10] As Susan Gal explains, the "antigenderism" discourses of the right function by "grafting on the authority of widely accepted, dominant moral values while also opposing them."[11] A lawsuit against the German car manufacturer Audi engaged this tactic when the complainant argued that the company's communications to employees, which used gender-inclusive terms, infringed upon his personal rights—a value often cited in arguments for gender-*inclusivity*.[12] In a similar way, the German far right has presented itself as a champion of women against the "threat" posed to them by Islam while also advocating for "classical"—that is, unequal—gender roles (I will discuss this more shortly). This rhetoric functions by asserting the naturalness of historical convention: the complementary model of the sexes is natural because it is traditional, traditional because it is natural.

Yet as any attentive student of history knows, there is nothing natural about the machinations that have been required to construct and prop up the "traditional" conception of the sexes. In language science and philosophy, the supposed naturalness of language's gender binarism was established only through the erection of complex, often illogical structures—such as the conflation of "women's languages" with "women's speech," or fanciful accounts of the reasoning behind various noun genders. The history of thinking about language and gender reveals that what has been claimed as natural and self-evident has, in fact, often been painstakingly manufactured. *Articulating Difference* has tried to bring this strangeness to the fore, showing the fantastical fictions produced and perpetuated in the service of language study. In particular, the book argued that sexual differentiation played a crucial epistemological role in the history of linguistic thought, transfiguring conjecture into truth, contingency into necessity. Between the late eighteenth and early twentieth centuries, claims about sexual complementarity formed an unquestioned ground for language science and philosophy's more speculative pursuits.

In the early text "Über Denken und Sprechen" ("On Thinking and Speaking"), Wilhelm von Humboldt, taking his cue from Herder, writes that "the Nature of thinking consists in reflecting, that is, in the act by which the thinking subject differentiates itself from its thought. In order to reflect, the mind has to arrest its continuous activity, grasp as a unit that which just presented itself and thus set it as an object over and against itself." This designation of thinking into units, Humboldt continues, is what we call *language*; language begins "immediately and simultaneously with the first act of reflexion."[13] The lesson of Humboldt's text is that there is no articulation without differentiation. The twin phenomena of language and thought are possible only within a being who recognizes the difference between his thought and his self, his self and his surroundings. The relationship between articulation and differentiation is not only a founding claim of such linguistic texts but, as I have maintained, a relationship that the texts themselves perform. For Humboldt and his contemporaries, articulating the connection *between* articulation and differentiation—that is, writing the scene of language's origin—was possible only if another difference, the difference of the sexes, was asserted first.

What does it mean to resurrect such claims today—and not in linguistic disciplines but in political discourse? To think further about the current conservative nostalgia for the understanding of language and sex prevalent in the nineteenth century, I would like to turn to a particularly revealing example, the 2016 "manifesto" of the German far-right populist party *Alternative für Deutschland* (Alternative for Germany), or AfD.

As of this writing (February 2024), the AfD is polling in second place nationwide in Germany ahead of elections later this year.[14] First formed in 2013 as a party centered on economic issues—chiefly "opposition to the Euro and rejection of all bailouts for member states of the Eurozone"[15]—the AfD's platform has developed into one of ethnonationalism.[16] In its "Manifesto for Germany," the AfD calls for a number of measures typical of ethnonationalist parties, including the restriction of immigration, the strengthening of law enforcement, a renewed commitment to the "traditional family"[17] and

Christianity,[18] and the legislation of German as the national language. "Our culture," the manifesto insists, "is inextricably linked to the German language, which has developed over centuries and which in itself is a reflection of its intellectual history, national identity within central Europe, and German set of basic values, which have all changed over time but have retained a unique core inventory. The bond which language creates amongst people should be maintained and protected."[19] To be sure, such a nationalist valorization of monolingualism is not new. The posited connection between language, culture, and nationhood has taken many forms but reaches a certain crystallization in the nineteenth century.[20] Claus Ahlzweig writes that though the term *Muttersprache,* or mother tongue, existed at least as far back as the twelfth century—in which it was used to distinguish the *Volkssprache,* or language of the people, from Latin, the language of scholars—it is not until the 1800s that it is invoked as part of an ideology of statehood and considered necessary for constructing national subjects.[21] We saw Fichte, for example, discuss monolingualism in this manner in his *Addresses to the German Nation.* Several years after the *Addresses,* Friedrich Ludwig Jahn, the "father" of German gymnastics and a militant language purist, would present an even more extreme version of Fichte's gendered linguistic nationalism. As Katarzyna Jastal illuminates, Jahn imagines the German language as a strong and powerful male body.[22] German, Jahn declares, asserts "its original right as original language in victorious war," while foreign words, which possess no "power of procreation" (*Zeugungskraft*), threaten to "emasculate" the German language—as well as, presumably, the men who speak it.[23]

Whereas French was the enemy of Fichte's address, written as it was during the Napoleonic wars, English is the main target of the AfD's ire, along with the languages spoken by recent refugees and immigrants to Germany. Yet this is not the only locus of the perceived threat. For, according to the AfD, German is endangered not only because of immigration and the global rise of English but also because German has become increasingly "gegendert" ("genderised").[24]

Gegendert is a strange term, and the manifesto does not provide a definition, although it most certainly refers to recent efforts to make languages better reflect the gender identities of their speakers, such as introducing gender-neutral and gender-inclusive pronouns and creating new words to describe gender and sexuality in flexible ways. These innovations come out of feminist, queer, and trans communities and, in the case of German, include practices such as the "gender gap" mentioned earlier,[25] as well as adding the letter x[26] to pronouns to render them gender neutral. As has occurred with English, German speakers seeking inclusive options have also created new pronouns.

The writer and artist Illi Anna Heger, for instance, has suggested the pronoun *xier*, showing how it works in different grammatical contexts.[27] Lann Horn-scheidt, whom I discussed briefly in chapter 3, has also written extensively on nonbinary and trans-inclusive language in German and Swedish.[28] Out of the 2007 Trans*Tagung in Berlin developed one of the earliest gender-neutral proposals in German: the "Sylvain Conventions," which includes the personal pronoun *nin* as well as the introduction of a new category of grammatical gender, the "indefinite."[29] Such reforms are presumably what the AfD has in mind to when it proclaims that it "firmly reject[s] politically correct language guidelines."[30] Party leaders have decried such changes, such as when one politician mocked different terms for gender identities in his "welcome" remarks during a meeting of the parliament of the state of Brandenburg.[31] Elsewhere in the manifesto, the AfD proclaims that gender-neutral words are a "violation of the naturally-developed culture and traditions of our language."[32]

As an English-derived term, *gegendert* signals that the "genderising" of language comes from the United States, where, according to the German conservative press, feminism and political correctness run amok.[33] The "genderizing" of language—and the so-called "ideology" of gender that it apparently relies upon—are thus presented as inauthentic to German culture. In addition to having a foreign word as its root, *gegendert* is also a verb in the form of a past participle. "Gegenderte Sprache" is language that *has been genderised*, externally acted upon. The AfD describes "gender" through a rhetoric of invasion similar to the derogatory way that it discusses Muslim immigrants—whom it calls "a danger to our state, our society, and our values."[34]

Nevertheless, the term *gegendert* also signifies something beyond "imported" language practices and the introduction of a few new pronouns and noun endings. Indeed, there exist in German established ways to refer to the language reforms just mentioned—*geschlechtergerechte Sprache* (gender-equal language) or *geschlechterneutrale Sprache* (gender-neutral language). It is striking that the manifesto instead chooses what was, in 2016, the less-established word *gegendert*. This term implies something more expansive, suggesting that the threat is not restricted to specific words, but that the introduction of these words will lead the German language itself to *have a gender* or *take on a gender*.

That the "genderising" of German is a problem presupposes that it is normally ungendered. Or we might say, more precisely, it presupposes that the German language has the *right* gender, the *unmarked* gender: the masculine. The similarities between the AfD's invective against "genderised" language and the nineteenth-century alignment of language and masculinity are clear. As we have seen, in the late eighteenth through the early twentieth century, "masculine" language was synonymous with "regular" language, just as the male human was synonymous with the human in general. Sexual difference is a difference belonging to woman alone. This is a perspective for which the AfD is plainly nostalgic. The party supports what it calls the "classical understanding of the roles of males and females,"[35] and furthermore declares that "gender ideology" (i.e., any understanding of gender that does not advance a complementary model of the sexes) "denies or marginalises natural differences between the sexes, denies traditional values, and specific roles."[36]

The strange reversal performed in conservative discourse, whereby making language gender *neutral* is referred to by its critics as "genderising," bespeaks a fantasy of masculine authority and feminine submission. Like Herder, Grimm and many other scholars examined in this book, the AfD assumes that language creation is the jurisdiction of the masculine alone. "Genderised" language poses a problem not only because of the way it changes the German language, and thus German cultural heritage, but also because of *who* is attempting to enact these changes. In the patriarchal conception of language, only men, not women or nonbinary people, have the authority for linguistic innovation.

The derision voiced through the term *gegenderte Sprache* is one way that the AfD would reassert the gender hierarchy it views as threatened. Faced with the increasing differentiation of gender, with a multiplicity of genders, the manifesto attempts, through the category "genderised," to impose a new dichotomy: there is language that has gender, and language that does not. This performative gesture faces a challenge on several fronts. First, in the twenty-first century, the term *gender* already refers not only to masculine and feminine but to a multiplicity of identities—to cisgender and transgender, to nonbinary and genderqueer and genderfluid, to name just a few. That the AfD invokes the so-called "ideology" of nonnormative genders, even if it is only to dismiss them, is a testament to their ubiquity and significance. As reactionary in its discourse as in its political platform, the AfD can only respond to a shift that has already taken place, telling of a new conception of gender in spite of itself. The second challenge facing the AfD's claims, of course, is that the pure, "ungenderised" language to which they aim to return has never existed. The modern conception of language that establishes an intimate bond between language, nation, and *Volk*—a conception that today's conservatives seek to resurrect—was already "genderised" in the nineteenth century, forged through ideas about masculinity, femininity, and their complementary relation.

Acknowledgments

The idea for this project first developed in the Department of Germanic Languages at Columbia University. I am grateful to Dorothea von Mücke, whose insightful questions helped shape the study and pushed me to develop my own voice. This book could not have been written without the support of Oliver Simons. Long after discharging his duties as PhD advisor, he has remained a generous interlocutor who has taught me so much about how to think and write.

At the University of Chicago, the Department of Germanic Studies has provided a supportive intellectual environment in which to pursue my idiosyncratic interests. David Levin has been an indefatigable mentor and source of new ideas. Conversations with Catriona MacLeod and Christopher Wild were especially formative for my thinking about this book, and I am grateful for their engagement with my work. Margareta Ingrid Christian has given me the best advice at every stage. I would also like to thank Florian Klinger, Eric Santner, and David Wellbery for their insights over the past half decade. I am grateful, too, to Daisy Delogu for sharing her wisdom. At Chicago, I was lucky to work with wonderful research assistants. I thank Siena Fite, Tae Ho Kim, and Nat Modlin for their care and interest in the project, and Amine Bouhayat for help with the French translations. Thank you as well to Brian Vetruba, Germanics librarian, for assistance in tracking down difficult sources, and to Emily Anderson for making the department run smoothly.

A research leave from the University of Chicago Humanities Division and a National Endowment for the Humanities Fellowship at the Newberry Library gave me access to important materials and time to craft the manuscript. This fellowship would not have been possible without the support of Yuliya

Komska and Imke Meyer. I thank them as well as the fellows in my cohort and, especially, Keelin Burke, Madeline Crispell, and Laura McEnany, whose comradery and encouragement made the year a joy.

At the University of Chicago Press, I have had the great fortune to work with Karen Merikangas Darling, whose guidance made the manuscript clearer and better argued. I would also like to thank the two anonymous readers for the Press, who read my work with care and whose suggestions greatly improved the book.

A number of friends and colleagues offered feedback on parts of the book at crucial moments. For their ideas and suggestions, I am grateful to Peter Erickson, Jason Farr, Alice Goff, Horst Lange, Trevor Perri, Kelly Wisecup, Lynn Wolff, and Tristram Wolff. Thank you also to Karin Nisenbaum for illuminating conversations about Fichte. I have been lucky to benefit from the support of several different communities in Chicago. My collaborators at the Newberry German Studies Scholarly Seminar—Alice Goff, Imke Meyer, Heidi Schlipphacke, Anna Souchuk, Lauren Stokes, and Erica Weitzman—have been a source of inspiration. My wonderful writing group, Tina Post, Leslie Rogers, Victoria Saramago, and Kathryn Takabvirwa, provided the motivation to draft every chapter. I thank Alice Goff and Ariana Strahl and Xan Holt and Kate Kelley for their friendship and love of Lake Michigan beaches.

It is hard to express the debt of gratitude that I owe my family. Robert and Lynn Alexander steadfastly encouraged me to pursue a PhD in German even when that seemed like the most impractical of life choices. Gretchen Alexander is always teaching me something new. My mother, Helene Alexander, has been a source of unending support and encouragement. Arthur Salvo is the best partner I could ask for, and his humor, enthusiasm, and keen analytical eye have greatly improved both my life and this book. I am grateful to him and Peter for creating our own version of the "household economy of the human species."

An earlier version of part of chapter 3 appeared in *MLN* as "The Sex of Language: Jacob Grimm on Grammatical Gender," vol. 136, no. 3 (April 2021): 770–93, © 2021 by Johns Hopkins University Press.

Notes

Introduction

1. In the German context, the two most well-known studies of women writers are perhaps Ruth-Ellen Boetcher Joeres's *Respectability and Deviance: Nineteenth-Century German Women Writers and the Ambiguity of Representation* (Chicago: University of Chicago Press, 1998) and *In the Shadow of Olympus: German Women Writers around 1800*, ed. Katherine R. Goodman and Edith Waldstein (Albany: State University of New York Press, 1992), although many more have been produced in the decades following. Most recently, Nicole Seifert has considered the historical denigration of women's writing in *Frauen Literatur: Abgewertet, vergessen, wiederentdeckt* (Cologne: Kiepenheuer & Witsch, 2021). On the etymologies and orthographies of words related to sex and sexuality, as well as the role of women in lexicography, see Jeffrey Masten, *Queer Philologies: Sex, Language, and Affect in Shakespeare's Time* (Philadelphia: University of Pennsylvania Press, 2016); Giovanni Iamartino, "Words by Women, Words on Women in Samuel Johnson's *Dictionary of the English Language*," in *Adventuring in Dictionaries: New Studies in the History of Lexicography*, ed. John Considine (Cambridge: Cambridge Scholars, 2010), 94–125; and Lindsay Rose Russell, *Women and Dictionary-Making: Gender, Genre, and English Language Lexicography* (Cambridge: Cambridge University Press, 2018). On the history of gendered pronouns, see Dennis Baron, *What's Your Pronoun? Beyond He and She* (New York: Liveright, 2020). On the history of rhetoric, see Lily Tonger-Erk, *Actio: Körper und Geschlecht in der Rhetoriklehre* (Berlin: de Gruyter, 2012). On language pedagogy, see Friedrich Kittler, *Discourse Networks 1800/1900*, trans. Michael Metteer with Chris Cullens, foreword David E. Wellbery (Stanford, CA: Stanford University Press, 1990).

2. On the significance of German language science, see E. F. K. Koerner, "Jacob Grimm's Position in the Development of Linguistics as a Science," in *The Grimm Brothers and the Germanic Past*, ed. Elmer H. Antonsen, James Marchand, and Ladislav Zgusta (Amsterdam: John Benjamins, 1990), 7–23.

3. Examples include Heymann Steinthal, *Geschichte der Sprachwissenschaft bei den Griechen und Römern mit besonderer Rücksicht auf die Logik* (Berlin: Ferd. Dümmler, 1863); Theodor Benfey, *Geschichte der Sprachwissenschaft und der orientalischen Philologie in Deutschland seit dem Anfange des 19. Jahrhunderts mit einem Rückblick auf die früheren Zeiten* (Munich: J. G. Cotta, 1869); and Berthold Delbrück, *Einleitung in das Sprachstudium: Ein Beitrag zur Geschichte und Methodik der vergleichenden Sprachforschung* (Leipzig: Breitkopf & Härtel, 1880). For more

on the nineteenth-century historiography of language science, see Henry M. Hoenigswald, "Nineteenth-Century Linguistics on Itself," in *Studies in the History of Western Linguistics*, ed. Theodora Bynon and F. R. Palmer (Cambridge: Cambridge University Press, 1986), 172–88.

4. Noam Chomsky, *Cartesian Linguistics: A Chapter in the History of Rationalist Thought* (New York: Harper & Row, 1966).

5. E. F. K. Koerner, "Editorial: Purpose and Scope of *Historiographia Linguistica*," *Historiographia Linguistica* 1, no. 1 (1974): 1–10, here 4.

6. This is the case for the *History of the Language Sciences: An International Handbook on the Evolution of the Study of Language from the Beginnings to the Present*, ed. Sylvain Auroux, E. F. K. Koerner, Hans-Josef Niederehe, and Kees Versteegh (Berlin: de Gruyter, 2000) as well as the four-volume *History of Linguistics*, first published in 1994 but republished several times hence. *History of Linguistics*, ed. Giulio C. Lepschy, 4 vols. (London: Pearson Education, 1994).

7. Sarah Pourciau, *The Writing of Spirit: Soul, System, and the Roots of Language Science* (New York: Fordham University Press, 2017).

8. *Women in the History of Linguistics*, ed. Wendy Ayres-Bennett and Helena Sanson (Oxford: Oxford University Press, 2021) is a recently published, wide-ranging account.

9. Robert A. Nye, "How Sex Became Gender," *Psychoanalysis and History* 12, no. 2 (2010): 195–209.

10. Judith Butler, *Bodies that Matter: On the Discursive Limits of "Sex"* (London: Routledge, 1993), xi–xii.

11. E. F. K. Koerner, "Linguistics vs. Philology: Self-Definition of a Field or Rhetorical Stance?" *Language Sciences* 19, no. 2 (1997): 167–75, here 168.

12. Wilhelm von Humboldt, "Ueber das vergleichende Sprachstudium in Beziehung auf die verschiedenen Epochen der Sprachentwicklung," in *Werke*, ed. Andreas Flitner and Klaus Giel, 5 vols. (Darmstadt: Wissenschaftliche Buchgesellschaft, 2002), 3:1–24.

13. For more on the history of *linguistics* and related terms, see Sylvain Auroux, "The First Uses of the French Word 'Linguistique' (1812–1880)," in *Papers in the History of Linguistics: Proceedings of the Third International Conference on the History of the Language Sciences (ICHoLS III), Princeton, 19–23 August 1984*, ed. Hans Aarsleff, L. G. Kelly, and Hans-Josef Niederehe (Amsterdam: John Benjamins, 1987), 447–59, as well as Pierre Swiggers, "A Note on the History of the Term *Linguistics*. With a Letter from Peter Stephen Du Ponceau to Joseph von Hammer-Purgstall," *Beiträge zur Geschichte der Sprachwissenschaft* 6 (1996): 1–17.

14. Sheldon Pollock, "Introduction," in *World Philology*, ed. Sheldon Pollock, Benjamin A. Elman, and Ku-ming Kevin Chang (Cambridge, MA: Harvard University Press, 2015), 1–24, here 22.

15. Ku-ming Kevin Chang, "Philology or Linguistics? Transcontinental Responses," in *World Philology*, 311–31, here 318.

16. Koerner, "Linguistics vs. Philology," 169.

17. August Schleicher, "Linguistik und Philologie," in *Die Sprachen Europas in systematischer Uebersicht* (Bonn: H. B. König, 1850), 1–5, here 1.

18. Friedrich Schlegel, *Über die Sprache und Weisheit der Indier* [1808], in *Kritische Friedrich-Schlegel-Ausgabe*, ed. Ernst Behler with Jean-Jacques Anstett and Hans Eichner (Munich: Ferdinand Schöningh, 1958–91), 8:105–433, here 137.

19. Johann Gottlieb Fichte, "Von der Sprachfähigkeit und dem Ursprung der Sprache," *J. G. Fichte-Gesamtausgabe*, ed. Reinhard Lauth and Hans Jacob (Stuttgart: Friedrich Frommann, 1966), 3:97–127; "On the Linguistic Capacity and the Origin of Language," in Jere Paul Surber,

Language and German Idealism: Fichte's Linguistic Philosophy (Atlantic Highlands, NJ: Humanities Press, 1996), 117–45.

20. Anna Morpurgo Davies, *Nineteenth-Century Linguistics*, vol. 4 of *History of Linguistics*, ed. Giulio C. Lepschy (London: Routledge, 1998), 3.

21. For more on the history of the German research university, see the introduction and primary documents included in *The Rise of the Research University: A Sourcebook*, ed. Louis Menand, Paul Reitter, and Chad Wellmon (Chicago: University of Chicago Press, 2017).

22. Morpurgo Davies, *Nineteenth-Century Linguistics*, 7.

23. See Els Elffers, "The Rise of General Linguistics as an Academic Discipline: Georg von der Gabelentz as a Co-Founder," in *From Early Modern to Modern Disciplines*, vol. 2 of *The Making of the Humanities*, ed. Rens Bod, Jaap Maat, and Thijs Weststeijn (Amsterdam: Amsterdam University Press, 2012), 55–70.

24. Georg von der Gabelentz, "Begriff der menschlichen Sprache," in *Die Sprachwissenschaft: Ihre Aufgaben, Methoden, und bisherigen Ergebnisse*, ed. Manfred Ringmacher and James McElvenny (Berlin: Language Science Press, 2016), 2–5.

25. Morpurgo Davies, *Nineteenth-Century Linguistics*, 231–32.

26. Pourciau, *The Writing of Spirit*, 62.

27. Hermann Osthoff and Karl Brugmann, *Morphologische Untersuchungen auf dem Gebiete der indogermanischen Sprachen* (Leipzig: S. Hirzel, 1878), vii.

28. Pourciau, *The Writing of Spirit*, 65.

29. Morpurgo Davies, *Nineteenth-Century Linguistics*, 268.

30. For example, although Hans Aarsleff argues against employing "hagiography" in the historiography of language study, he makes a strong case for the influence of Condillac on the development of linguistic thought. Michael Forster, on the other hand, champions Herder. See Hans Aarsleff, *From Locke to Saussure: Essays on Study of Language and Intellectual History* (Minneapolis: University of Minnesota Press, 1985), and Michael N. Forster, *After Herder: Philosophy of Language in the German Tradition* (Oxford: Oxford University Press, 2010).

31. Michel Foucault, *The Order of Things: An Archaeology of the Human Sciences* (New York: Vintage Books, 2012), 290.

32. Johann Gottfried Herder, *Abhandlung über den Ursprung der Sprache*, in *Werke in zehn Bänden*, vol. 1, ed. Ulrich Gaier (Frankfurt a.M.: Deutscher Klassiker, 1985), 695–810, here 711; *Treatise on the Origin of Language*, in *Philosophical Writings*, trans. and ed. Michael N. Forster (Cambridge: Cambridge University Press, 2002), 65–164, here 112.

33. Jacob Grimm, *Über den Ursprung der Sprache* (Berlin: Ferd. Dümmler, 1866), 31; quoted and translated in Foucault, *Order of Things*, 317, translation modified.

34. See, for instance, Ian Hacking, "Night Thoughts on Philology," in *Historical Ontology* (Cambridge, MA: Harvard University Press, 2002), 140–51. Pourciau also argues against Foucault's understanding of the modern episteme as fundamentally renouncing the paradigm of representation. See *The Writing of Spirit*, 8–9.

35. See Patricia M. Mazón, *Gender and the Modern Research University: The Admission of Women to German Higher Education, 1865–1914* (Stanford, CA: Stanford University Press, 2003).

36. James C. Albisetti, "The German Ideal of Womanhood and Education," in *Schooling German Girls and Women: Secondary and Higher Education in the Nineteenth Century* (Princeton, NJ: Princeton University Press, 1988), 3–22.

37. Albisetti, *Schooling*, 20. Albisetti notes, for example, that "in Bavaria, an official decree of 1804 'stipulated that women should not study any sciences.'"

38. Medicine and teaching were some of the first disciplines to open up to women, while theology, law, and philosophy "remained a male stronghold." See Mazón, *Gender and the Modern Research University*, 15, and Albisetti, "The Decisive Reforms in Female Education," in *Schooling*, 238–73.

39. See Marcin Kilarski, *Nominal Classification: A History of Its Study from the Classical Period to the Present* (Amsterdam: John Benjamins, 2013), 83–97.

40. Stefani Engelstein, *Sibling Action: The Genealogical Structure of Modernity* (New York: Columbia University Press, 2017), 5.

41. Jocelyn Holland, *German Romanticism and Science: The Procreative Poetics of Goethe, Novalis, and Ritter* (New York: Routledge, 2009); Helmut Müller-Sievers, *Self-Generation: Biology, Philosophy, and Literature around 1800* (Stanford, CA: Stanford University Press, 1997). See also his *The Science of Literature: Essays on an Incalculable Difference*, trans. Chadwick Truscott, Paul Babinski, and Helmut Müller-Sievers (Berlin: de Gruyter, 2015).

42. Donna Haraway, *Primate Visions: Gender, Race, and Nature in the World of Modern Science* (New York: Routledge, 1989); Sarah S. Richardson, *Sex Itself: The Search for Male and Female in the Human Genome* (Chicago: University of Chicago Press, 2013); Anne Fausto-Sterling, *Sexing the Body: Gender Politics and the Construction of Sexuality* (New York: Basic Books, 2000).

43. Londa Schiebinger, *Nature's Body: Gender in the Making of Modern Science* (New Brunswick, NJ: Rutgers University Press, 2004), x.

44. Michèle Le Doeuff, *Hipparchia's Choice: An Essay Concerning Women, Philosophy, Etc.*, trans. Trista Selous (Oxford: Blackwell, 1991), 13.

45. See Mazón, *Gender and the Modern Research University*, 19.

46. Robin Lakoff's 1973 "Language and Women's Place" was influential in its argument that the language used both by and about women perpetuates women's inequality in the United States. Two years later, Mary Ritchie Key published *Male/Female Language*, which studied how the social structures of masculinity and femininity operate linguistically. In 1980, Dale Spender's *Man Made Language* argued that English remains "primarily under male control." In Germany, Luise F. Pusch proposed *feministische Linguistik* as a field concerned with investigating how patriarchal values inflect different languages. Similarly, Senta Trömel-Plötz argued that linguistics must pay attention to and study women's speech. More recently, linguists have studied the negative perceptions of "uptalk" and "vocal fry," intonations associated with femininity and effeminacy.

See Robin Lakoff, "Language and Women's Place," *Language and Society* 2, no. 1 (April 1973): 45–80; Mary Ritchie Key, *Male/Female Language: With a Comprehensive Bibliography* (Metuchen, NJ: Scarecrow Press, 1975); Dale Spender, *Man Made Language* (London: Routledge, 1980); Deborah Tannen, *You Just Don't Understand: Women and Men in Conversation* (New York: Ballantine Books, 1990); *Talking from 9 to 5: Women and Men at Work* (New York: Avon, 1994); Sally McConnell-Ginet, "Intonation in a Man's World," *Signs* 3, no. 3 (Spring 1978): 541–59; "Language and Gender," in *Linguistics: The Cambridge Survey*, ed. Frederick J. Newmeyer (Cambridge: Cambridge University Press, 1988), 4:75–99; with Penelope Eckert, "Think Practically and Look Locally: Language and Gender as Community-Based Practice," *Annual Review of Anthropology* 21 (1992): 461–90; Deborah Cameron, *Feminism and Linguistic Theory* (London: Palgrave Macmillan, 1985); *The Myth of Mars and Venus: Do Men and Women Really Speak Different Languages?* (Oxford: Oxford University Press, 2008); Luise F. Pusch, *Alle Menschen werden Schwestern* (Frankfurt a.M.: Suhrkamp, 1990); Senta Trömel-Plötz, *Frauensprache: Sprache der Veränderung* (Frankfurt a.M.: Fischer, 1982); Penelope Eckert and Sally McConnell-Ginet,

"Being Assertive . . . Or Not," in *Language and Gender* (Cambridge: Cambridge University Press, 2013), 141–63.

47. Robert Lawson, *Language and Mediated Masculinities: Cultures, Contexts, Constraints* (Oxford: Oxford University Press, 2023), 31–32.

48. While many feminist theories of this period originated in France, the discussion quickly became transnational. "Is the pen a metaphorical penis?" asked the American literary critics Sandra Gilbert and Susan Gubar in 1979, coining "anxiety of authorship" as the female alternative to Harold Bloom's "anxiety of influence" model of literary paternity. That same year, Elaine Showalter introduced the term *gynocriticism* into English as a way to describe feminist scholarship that focused on the woman writer. Similarly, Patricia Meyer Spacks's 1975 *The Female Imagination* studied women's writing to contend that "a special female self-awareness emerges through literature in every period." In works such as these, *women's language* appears not, or not merely, as a reference to the gender of specific authors, but as a statement about the gender of the texts they produce. Such criticism asserts, in other words, that there exist distinctly *feminine* texts, texts that reflect—diegetically, aesthetically, formally—distinctly feminine experiences and concerns.

See Sandra M. Gilbert and Susan Gubar, *The Madwoman in the Attic: The Woman Writer and the Nineteenth-Century Literary Imagination* (New Haven, CT: Yale University Press, 1979); Elaine Showalter, "Towards a Feminist Poetics," in *Women Writing and Writing about Women*, ed. Mary Jacobus (London: Croom Helm, 1979), 22–41; and Patricia Meyer Spacks, *The Female Imagination* (New York: Knopf, 1975).

49. Hélène Cixous, "Le Rire de la Méduse," *L'Arc* 61 (May 1975): 39–54, here 39; "The Laugh of the Medusa," trans. Keith Cohen and Paula Cohen, *Signs* 1, no. 4 (Summer 1976): 875–93, here 875.

50. Cixous, "Méduse," 39; "Medusa," 875.

51. Like *écriture féminine, parler femme* programmatically evades definition; Irigaray describes it as a kind of subversive mimesis of the master discourse. See the chapter "Questions" in her *This Sex Which is Not One*, trans. Catherine Porter with Carolyn Burke (Ithaca, NY: Cornell University Press, 1985), 119–69. Kristeva's semiotic is—in Elizabeth Grosz' summary—"a *presignifying energy*" associated with the maternal. See Julia Kristeva, *Revolution in Poetic Language* (New York: Columbia University Press, 1984), as well as Elizabeth Grosz, *Sexual Subversions: Three French Feminists* (Sydney: Allen & Unwin, 1989), 42.

52. See, for example, *Language and Liberation: Feminism, Philosophy, and Language*, ed. Christina Hendricks and Kelly Oliver (Albany: State University of New York Press, 1999); Dani Cavallaro, *French Feminist Theory: An Introduction* (London: Continuum, 2003); and *Contemporary French Feminism*, ed. Kelly Oliver and Lisa Walsh (Oxford: Oxford University Press, 2004).

53. Tristram Wolff, *Against the Uprooted Word: Giving Language Time in Transatlantic Romanticism* (Stanford, CA: Stanford University Press, 2022), 18 and 27.

54. Wilhelm Bleek, *Über den Ursprung der Sprache* (Cape Town: Van de Sandt de Villiers, 1867), xix; *On the Origin of Language*, ed. Ernst Haeckel, trans. Thomas Davidson (New York: L. W. Schmidt, 1869), xxvi.

55. Jacques Rancière, *The Names of History: On the Poetics of Knowledge*, trans. Hassan Melehy (Minneapolis: University of Minnesota Press, 1994), 8.

56. Joseph Vogl, "Für eine Poetologie des Wissens," in *Die Literatur und die Wissenschaften, 1770–1930*, ed. Karl Richter, Jörg Schönert, and Michael Titzmann (Stuttgart: M & P, 1997), 107–27, here 125.

57. Ruth Römer, *Sprachwissenschaft und Rassenideologie in Deutschland* (Munich: Wilhelm Fink, 1985); Maurice Olender, *The Languages of Paradise: Race, Religion, and Philology in the Nineteenth Century*, trans. Arthur Goldhammer (Cambridge, MA: Harvard University Press, 1992); Geoffrey Galt Harpham, "Roots, Races, and the Return to Philology," *Representations* 106, no. 1 (Spring 2009): 34–62; Joseph Errington, *Linguistics in a Colonial World: A Story of Language, Meaning, and Power* (Malden, MA: Blackwell, 2008); Tuska Benes, *In Babel's Shadow: Language, Philology, and the Nation in Nineteenth-Century Germany* (Detroit: Wayne State University Press, 2008); Markus Messling, "Text and Determination: On Racism in 19th Century European Philology," *Philological Encounters* 1 (2016): 79–104.

58. David Golumbia, "The Deconstruction of Philology," *boundary 2* 48, no. 1 (2021): 17–34, here 20.

59. Siraj Ahmed, *Archaeology of Babel: The Colonial Foundations of the Humanities* (Stanford, CA: Stanford University Press, 2018), 200.

60. Benes, *Babel's Shadow*, 200.

Chapter One

1. According to Greek mythology, a red rose grew from the tears that Aphrodite, the goddess of love, shed when her beloved Adonis was killed. *Encyclopedia of Greek and Roman Mythology*, ed. Luke Roman and Monica Roman (New York: Facts on File, 2010), 11.

2. On the publication history, see Jere Paul Surber's introduction to his translation of Fichte's text. Surber, *Language and German Idealism*, 119.

3. Fichte, "Ursprung," 97; "Linguistic Capacity," 119, translation modified.

4. Fichte, "Ursprung," 99; "Linguistic Capacity," 121.

5. Fichte, "Ursprung," 99; "Linguistic Capacity," 121.

6. What constitutes the first conjectural history is a point of scholarly contention. The term was introduced in 1793 by Dugald Stewart to describe Adam Smith's theory of the origin of language, but many critics argue that the form predates this coinage. Frank Palmeri contends that Bernard de Mandeville's 1729 *Fable of the Bees* is the first instance of conjectural history, although ancient writers also made use of constitutive aspects of the genre. Lydia Moland, on the other hand, argues that Thomas Hobbes's *Leviathan* "sets the modern stage" of conjectural histories, while Jenny Mander writes that Pierre de Marivaux's novel *Le Paysan parvenu* (1734–35) can be read as an "embryonic conjectural history." For the purposes of this chapter, I am less interested in the form's origin point than in how its components change over time. For more on the term *conjectural history* and *its* history, see Dugald Stewart, "Account of the Life and Writings of Adam Smith, LL.D.," in *The Collected Works of Dugald Stewart*, ed. William Hamilton (Edinburgh: Thomas Constable, 1858), 10:1–100; Frank Palmeri, *Stages of Nature, Stages of Society: Enlightenment Conjectural History and Modern Social Discourse* (New York: Columbia University Press, 2016), 27; Lydia L. Moland, "Conjectural Truths: Kant and Schiller on Educating Humanity," in *Kant and His German Contemporaries*, ed. Daniel O. Dahlstrom (Cambridge: Cambridge University Press, 2018), 2:91–108, here 92; and Jenny Mander, "Marivaux's *Paysan parvenu* and the Genre of Conjectural History," *Nottingham French Studies* 48, no. 3 (Autumn 2009): 20–30, here 23.

7. Jean-Jacques Rousseau, *Discours sur l'origine et les fondements de l'inégalité parmi les hommes* (Paris: Gallimard, 2006), 31; *Discourse on the Origin of Inequality*, ed. Patrick Coleman, trans. Franklin Philip (Oxford: Oxford University Press, 1994), 24.

8. Friedrich Schiller, "Ueber die ästhetische Erziehung des Menschen in einer Reihe von Briefen," in *Philosophische Schriften I*, ed. Benno von Wiese, vol. 20 of *Schillers Werke: Nationalausgabe* (Weimar: Hermann Böhlaus Nachfolger, 1962), 309–412, here 389; quoted in Moland, "Conjectural Truths," 108.

9. This occurs in Herder's text, for example. See Herder, *Abhandlung*, 723–24; *Treatise*, 88–9.

10. Herder, *Abhandlung*, 810; *Treatise*, 164.

11. Immanuel Kant, "Mutmaßlicher Anfang der Menschengeschichte," in *Gesammelte Schriften (Akademie-Ausgabe): Electronic Edition*, vol. 8, *Abhandlungen nach 1781* (Charlottesville, VA: InteLex, 1999), 110–23, here 110; "Conjectural Beginning of Human History," trans. Allen A. Wood, in *Anthropology, History, and Education*, ed. Günter Zöller and Robert B. Louden (Cambridge: Cambridge University Press, 2007), 163–75, here 163.

12. Palmeri argues that "must have," which he calls the "conjectural necessity," is constitutive of the genre (*Stages of Nature*, 16). Many eighteenth-century conjectural histories, however, actually use "may" or "might have," marking their narrative as hypothetical. Dugald Stewart, for instance, writes in his biography of Adam Smith that "when we cannot trace the process by which an event *has been* produced, it is often of importance to be able to show how it *may have been* produced by natural causes" ("Adam Smith," 34).

13. Jean Frain du Tremblay, *Traité des langues* [. . .] (Paris: Jean-Baptiste Delespine, 1703), 29; *A Treatise of Languages* [. . .], trans. M. Halpenn (London: D. Leach, 1725), 19.

14. Frain du Tremblay, *Traité des Langues*, 31; *Treatise of Languages*, 20.

15. Frain du Tremblay, *Traité des Langues*, 32; *Treatise of Languages*, 20–21.

16. Warburton asks: "Can any thing be more familiar than the Image these Words give one of a Learner of his Rudiments?" William Warburton, *The Divine Legation of Moses Demonstrated*, 2 vols. (London: Fletcher Gyles, 1741), 2:82.

17. Warburton, *Divine Legation of Moses*, 2:81.

18. Jean Henri Samuel Formey, "Réunion des principaux moyens employés pour découvrir l'origine du langage, des idées, et des connais," in *Anti-Emile* (Berlin: Joachim Pauli, 1763), 211–30, here 221.

19. Formey, "Réunion," 221.

20. Formey, "Réunion," 225.

21. Louis de Bonald, for example, a philosopher and opponent of the French Revolution, defends the divine origin of language in an 1818 text. See de Bonald, "De l'origine du langage," in his *Récherches philosophiques sur les premiers objets des connaissances morales* (Paris: A. Le Clere, 1882), 61–117.

22. Other divine origin theories take different approaches. Johann Peter Süssmilch's *Versuch* makes an argument based on logic. To invent language, he asserts, humans would have to already have an idea of language—but without language, they can have no ideas. The only way to resolve this paradox, Süssmilch contends, is for language to have been a divine creation. Johann Peter Süssmilch, *Versuch eines Beweises, dass die erste Sprache ihren Ursprung nicht von Menschen, sondern allein vom Schöpfer erhalten habe* (Berlin: Buchladen der Realschule, 1766).

23. Hans Aarsleff, "The Rise and Decline of Adam and his *Ursprache* in Seventeenth-Century Thought," in *The Language of Adam/Die Sprache Adams*, ed. Allison P. Coudert (Wiesbaden: Harrassowitz, 1999), 277–95, here 277.

24. John Locke, *An Essay Concerning Human Understanding*, ed. Peter H. Nidditch (Oxford: Clarendon Press, 1990), 466.

25. Genesis 4:19.

26. Locke, *Human Understanding*, 466.

27. Robert Filmer, *Patriarcha, or the Natural Power of Kings* (London: R. Chiswell, 1680).

28. John Locke, *First Treatise*, in *Two Treatises of Government* and *A Letter Concerning Toleration*, ed. Ian Shapiro with essays by John Dunn, Ruth W. Grant, and Ian Shapiro (New Haven, CT: Yale University Press, 2003), 7–99, here 23.

29. Locke, *Human Understanding*, 33.

30. As Grant further argues, for Locke, man's authority over his wife in marriage is rooted in custom rather than in moral law: "Association between a man and a woman in marriage, which Locke calls 'conjugal society,' is formed by voluntary contract for the purpose of producing and rearing children. [. . .] Since unanimity of judgment cannot always be expected, the final determination must be lodged in one or the other of them in cases of conflict. This conjugal power is 'naturally' lodged in the husband as the 'abler and stronger,' and Locke remarks, 'there is, I grant, a foundation in nature for it,' though it is a customary or conventional arrangement." Ruth W. Grant, "John Locke on Women and the Family," in John Locke, *Two Treatises of Government* and *A Letter Concerning Toleration*, ed. Ian Shapiro (New Haven, CT: Yale University Press, 2003), 286–308, here 291–92.

31. See Grant, "Locke on Women and the Family," 292.

32. Bernard de Mandeville, *The Fable of the Bees: or, Private Vices, Publick Benefits*, ed. F. B. Kaye (Oxford: Oxford University Press, 2014), 2:287–88.

33. Johann David Michaelis, "Preisschrift vom Ursprung der Sprache, so ich am 12. Dec. 1770 an die Akademie zu Berlin gesandt habe" (unpublished manuscript, Codex Michaelis 72), Niedersächsisches Staats- und Universitätsbibliothek Göttingen Abteilung Handschriften und Alte Drucke Nachlass Michaelis; quoted in Cordula Neis, *Anthropologie im Sprachdenken des 18. Jahrhunderts: Die Berliner Preisfrage nach dem Ursprung der Sprache* (Berlin: de Gruyter, 2003), 534. Michaelis entered this essay into the Berlin Academy competition that Herder famously won. In an earlier text, submitted to a different prize competition in 1759, Michaelis had furthermore claimed that women play an important part in shaping the words of a language. See Michaelis's *Dissertation [. . .] sur l'influence réciproque du langage sur les opinions, et les opinions sur le langage* (Berlin: Haude and Spener, 1760), 78. For this reference I am indebted to Avi Lifschitz, *Language and Enlightenment: The Berlin Debates of the Eighteenth Century* (Oxford: Oxford University Press, 2016), 128.

34. Michaelis, "Preisschrift" (Codex Michaelis 72), quoted in Neis, *Anthropologie*, 534.

35. Étienne Bonnot de Condillac, *Essai sur l'origine des connaissances humaines*, vol. 1 of *Ouevres complètes de Condillac* (Paris: Ch. Houel, 1798), 257; *An Essay on the Origin of Human Knowledge*, trans. Thomas Nugent (London: J. Nourse, 1756), 169.

36. Condillac, *Essai sur l'origine des connaissances humaines*, 262; *Origin of Human Knowledge*, 173.

37. Condillac, *Essai sur l'origine des connaissances humaines*, 261; *Origin of Human Knowledge*, 172, translation modified. In a footnote, Condillac further explains that the two children are the most "natural choice": he used them as his example "because I did not think it sufficient for a philosopher to say a thing was effected by extraordinary means, but judged it to be also incumbent upon him to explain how it could have happened according to the ordinary course of nature." Condillac, *Essai sur l'origine des connaissances humaines*, 259; *Origin of Human Knowledge*, 171.

38. Condillac, *Essai sur l'origine des connaissances humaines*, 265; *Origin of Human Knowledge*, 175.

39. Condillac, *Essai sur l'origine des connaissances humaines*, 260; *Origin of Human Knowledge*, 171.

40. Abbé Copineau, *Essai synthétique sur l'origine et la formation des langues* (Paris: Ruault, 1774), 8.

41. Jean-Jacques Rousseau, *Essai sur l'origine des langues, où il est parlé de la mélodie et de l'imitation musicale*, ed. Charles Porset (Bordeaux: Ducros, 1970), 27; "Essay on the Origin of Language," in Jean-Jacques Rousseau and Johann Gottfried Herder, *On the Origin of Language*, trans. John H. Moran and Alexander Gode (Chicago: University of Chicago Press, 1986), 5–84, here 5.

42. Rousseau, *Essai sur l'origine des langues*, 27; "Origin of Language," 5.

43. "It is neither hunger nor thirst but love, hatred, pity, anger, which drew from them the first words. Fruit does not disappear from our hands. One can take nourishment without speaking. One stalks in silence the prey on which one would feast. But for moving a young heart, or repelling an unjust aggressor, nature dictates accents, cries, lamentations. There we have the invention of the most ancient words; and that is why the first languages were singable and passionate before they became simple and methodical." Rousseau, *Essai sur l'origine des langues*, 43; "Origin of Language," 12.

44. Rousseau, *Essai sur l'origine des langues*, 29; "Origin of Language," 6.

45. Jean-Jacques Rousseau, *Émile, ou de l'éducation* (Paris: Garnier-Flammarion, 1966), 507; *Emile, or On Education*, trans. Allan Bloom (New York: Basic Books, 1979), 386.

46. Penny A. Weiss, *Gendered Community: Rousseau, Sex, and Politics* (New York: New York University Press, 1993), 40.

47. Rousseau, *Émile*, 507; *Emile*, 387.

48. Adam Smith, "Considerations Concerning the Formation of Languages," in *The Works of Adam Smith*, ed. Dugald Stewart (London: T. Cadell, 1811), 5:1–48, here 3.

49. See Chris Nyland, "Adam Smith, Stage Theory, and the Status of Women," in *The Status of Women in Classical Economic Thought*, ed. Robert Dimand and Chris Nyland (Northampton, MA: Edward Elgar, 2003), 86–107. Nyland shows how Smith borrowed much from Montesquieu without importing Montesquieu's "claim that females had less innate intellectual capacity" (87).

50. Lord James Burnett Monboddo, *Of the Origin and Progress of Language*, vol. 1, 2nd ed. (Edinburgh: J. Balfour, 1774), 481.

51. Monboddo, *Origin and Progress of Language*, 475.

52. See Sebastian Domsch, "Language and the Edges of Humanity: Orang-Utans and Wild Girls in Monboddo and Peacock," *Zeitschrift für Anglistik und Amerikanistik* 56, no. 1 (January 2008): 1–11.

53. Monboddo, *Origin and Progress of Language*, 475–76.

54. Annette F. Timm and Joshua A. Sanborn, *Gender, Sex and the Shaping of Modern Europe: A History from the French Revolution to the Present Day*, 3rd ed. (London: Bloomsbury, 2022), 35.

55. Margaret R. Hunt, *Women in Eighteenth-Century Europe* (Harlow: Pearson Longman, 2010), 14.

56. Karen Offen, *European Feminisms 1700–1950: A Political History* (Stanford, CA: Stanford University Press, 2000), 31–49.

57. Lieselotte Steinbrügge, *The Moral Sex: Woman's Nature in the French Enlightenment*, trans. Pamela E. Selwyn (Oxford: Oxford University Press, 1995), 6.

58. Isabel V. Hull, *Sexuality, State, and Civil Society in Germany, 1700–1815* (Ithaca, NY: Cornell University Press, 1996), 211.

59. Schiebinger, *Nature's Body*, 55.

60. Schiebinger, *Nature's Body*, 55.

61. Wilhelm von Humboldt, *Ueber die Verschiedenheiten des menschlichen Sprachbaues und ihren Einfluss auf die geistige Entwicklung des Menschengeschlechts*, in *Werke in fünf Bänden* (Darmstadt: Wissenschaftliche Buchgesellschaft, 1963), 3:368–756, here 435; *On the Diversity of Human Language Construction and Its Influence on the Mental Development of the Human Species*, ed. Michael Losonsky, trans. Peter Heath (Cambridge: Cambridge University Press, 1999), 60.

62. Herder, *Abhandlung*, 787; *Treatise*, 142.

63. Herder, *Abhandlung*, 787; *Treatise*, 142.

64. For example, this is the approach of the recent *Herder: Philosophy and Anthropology*, ed. Anik Waldow and Nigel DeSouza (Oxford: Oxford University Press, 2017), 182.

65. See, for example, Linda Dietrick, "Herder as a Mentor of Literary Women," *Herder Jahrbuch* 14, ed. Rainer Godel and Johannes Schmidt (Heidelberg: Synchron, 2018), 119–43.

66. Herder, *Abhandlung*, 797; *Treatise*, 153

67. Johann Gottfried Herder, *Ideen zur Philosophie der Geschichte der Menschheit*, in *Werke in zehn Bänden*, vol. 6, ed. Martin Bollacher (Frankfurt a.M.: Deutscher Klassiker, 1989); *Outlines of a Philosophy of the History of Man*, vol. 1, 2nd ed., trans. T. Churchill (London: Luke Hansard, 1803).

68. Jennifer Fox, "The Creator Gods: Romantic Nationalism and the En-genderment of Women in Folklore," *Journal of American Folklore* 100, no. 398 (October–December 1987): 563–72, here 567.

69. Herder, *Abhandlung*, 784; *Treatise*, 140.

70. Herder, *Ideen*, 158; *Outlines*, 179, translation modified.

71. See, for example, Christoph F. E. Holzhey, "On the Emergence of Sexual Difference in the 18th Century: Economies of Pleasure in Herder's *Liebe und Selbstheit*," *German Quarterly* 79, no. 1 (Winter 2006): 1–27. Johann Gottfried Herder, "Liebe und Selbstheit. Ein Nachtrag zum Briefe des Herrn Hemsterhuis über das Verlangen," in *Werke in zehn Bänden*, vol. 4, ed. Jürgen Brummack and Martin Bollacher (Frankfurt a.M.: Deutscher Klassiker, 1994), 405–24, especially 421.

72. Herder, *Ideen*, 322; *Outlines*, 386.

73. Herder, *Abhandlung*, 801; *Treatise*, 156.

74. Herder, *Abhandlung*, 787–88; *Treatise*, 143.

75. Benedict Anderson, *Imagined Communities: Reflections on the Origin and Spread of Nationalism* (London: Verso, 2006), 68.

76. "VATERSCHWEISZ, m.," in *Deutsches Wörterbuch von Jacob Grimm und Wilhelm Grimm*, digital edition from the Trier Center for Digital Humanities, Version 01/21. The source for this entry is a text by Friedrich (Maler) Müller from 1803.

77. Herder, *Abhandlung*, 788; *Treatise*, 144–45.

78. "All the objects of our senses become ours only to the extent that we *become aware* of them; that is, we designate them, in a more or less clear and vivid fashion, with the stamp *of our consciousness*." Herder, "Über Bild, Dichtung und Fabel," in *Werke in zehn Bänden*, vol. 4, ed. Jürgen Brummack and Martin Bollacher (Frankfurt a.M.: Deutscher Klassiker), 631–77, here 635; "On Image, Poetry, and Fable," in *Selected Writings on Aesthetics*, trans. and ed. Gregory Moore (Princeton, NJ: Princeton University Press, 2006), 357–82, here 358. For more on Herder's theory of sensuous perception, and his use of "stamp" and "seal" metaphors, see Amanda Jo Goldstein, *Sweet Science: Romantic Materialism and the New Logics of Life* (Chicago: University of Chicago Press, 2017), 28.

NOTES TO PAGES 30-33

79. See Johann Gottfried Herder, "Vom Erkennen und Empfinden der menschlichen Seele," in *Werke in zehn Bänden*, 4:357–58; "On the Cognition and Sensation of the Human Soul," in *Philosophical Writings*, trans. and ed. Michael N. Forster (Cambridge: Cambridge University Press, 2002), 187–244, here 211.

80. This is Jürgen Habermas's description of Herder. See Habermas, *Der philosophische Diskurs der Moderne: Zwölf Vorlesungen* (Frankfurt a.M.: Suhrkamp, 1985), 80.

81. Herder, *Ideen*, 324; *Outlines*, 389, translation modified.

82. Herder, *Ideen*, 339; *Outlines*, 410.

83. Herder, *Ideen*, 342; *Outlines*, 413.

84. "Like the human, the ape has no determinate instinct—its mode of thinking stands close on the brink of reason, the brink of imitation. It imitates everything, and therefore its brain must be fitted for a thousand combinations of sensitive ideas, of which no other brute is capable: for neither the wise elephant, nor the sagacious dog, is capable of doing what the ape can perform: *it would perfect itself*. But this it cannot: the door is shut against it: its brain is incapable of combining with its own ideas those of others, and making what it imitates as it were its own." Herder, *Ideen*, 116–17; *Outlines*, 126, translation modified.

85. Herder, *Ideen*, 357–58; *Outlines*, 433, my emphasis, translation modified.

86. Herder, *Ideen*, 339; *Outlines*, 409–10.

87. Herder, *Ideen*, 324; *Outlines*, 389, translation modified.

88. References to daughters appear only sparingly. In the *Treatise*, for instance, while discussing the "chain of instruction" from parents to children, Herder writes that "each individual is a son or daughter, was educated [*gebildet*] through instruction" (155). Yet as we saw in our analysis of Part II, whenever Herder narrates an example of language instruction, he does so through the example of fathers and sons.

89. Here, Herder deploys the popular trope that a culture's treatment of women, influenced by its climate, is an index of its intelligence and civilization. Unsurprisingly, Germans come out on top: not only the modern German, but also the ancient German, who, "in his wild forests, understood the worth of the female sex, and enjoyed in them the noblest qualities of man, fidelity, prudence, courage, and chastity" (*Outlines*, 383). Compare this to the treatment of women in "eastern regions," where the warm climate induces an early awakening of sexual desire, and causes man to "abuse the superiority of his sex, and endeavor to form a garden of these perishable flowers"—in other words, a harem.

For a historical overview of the claim that women act as a "barometer" of civilization, see Karen O'Brien, "From Savage to Scotswoman: The History of Femininity," in *Women and the Enlightenment in Eighteenth-Century Britain* (Cambridge: Cambridge University Press, 2009), 68–109.

90. Herder, *Ideen*, 320–21; *Outlines*, 384–85.

91. Herder, *Ideen*, 322; *Outlines*, 386.

92. Herder, *Ideen*, 321; *Outlines*, 385.

93. Herder, *Ideen*, 316; *Outlines*, 378.

94. According to Greek mythology, the apple of discord is the golden apple thrown by Eris at the wedding of Peleus and Thetis. Intended for the most beautiful goddess, it effects a competition among Aphrodite, Hera, and Athena. Paris, forced to judge, chooses Aphrodite in return for Helen, and thereby precipitates the Trojan War. See Roman and Roman, *Encyclopedia*, 389. Viewed from a different angle, the apple is the fruit of the tree of knowledge, which, through Eve, brings conflict into the Garden of Eden. Among her punishments is God's decree that "your

desire will be for your husband, and he shall rule over you" (Gen. 3:16 [New Revised Standard Version]).

95. John Milton, *Paradise Lost*, ed. Gordon Teskey (New York: W. W. Norton, 2005), bk. 4, p. 91, lines 487-91.

96. Johann Gottfried Herder, *Herders Briefwechsel mit Caroline Flachsland*, ed. Hans Schauer (Weimar: Goethe-Gesellschaft, 1926), 1:18; quoted in Birgit Nübel, "Krähende Hühner und gelehrte Weiber: Aspekte des Frauenbildes bei Johann Gottfried Herder," in *Herder Jahrbuch*, ed. Wulf Koepke and Wilfried Malsch (Stuttgart: J. B. Metzler, 1994), 29-49, here 48.

97. Johann Gottfried Herder, "Über die neuere deutsche Literatur. Fragmente, als Beilagen zu den Briefen, die neueste Literatur betreffend. Dritte Sammlung. 1767," in *Werke in zehn Bänden*, ed. Ulrich Gaier (Frankfurt a.M.: Deutscher Klassiker, 1985), 1:367-539, here 396; quoted in Nübel, "Krähende Hühner," 39.

98. This is from a letter to Caroline Flachsland written on September 20, 1770. See Herder, *Briefwechsel mit Caroline Flachsland*, ed. Hans Schauer (Weimar: Goethegesellschaft, 1926-28), 1:47. Also quoted in Nübel, "Krähende Hühner," 48, my emphasis.

99. Herder, *Briefwechsel*, 1:47-48.

100. Herder, *Treatise*, 82.

101. Herder, *Briefwechsel*, 1:46. Also quoted in Nübel, "Krähende Hühner," 49. In *How Philosophy Can Become More Universal*, Herder writes that women can be improved through education, but they should not learn the same things as men; philosophy and most sciences are off limits, as are politics and war, and anything that would render women learned. Instead, women should "feel virtue" and "think beautifully" as they learn "society and taste." See Herder, "How Philosophy Can Become More Universal and Useful for the Benefit of the People (1765)," in *Philosophical Writings*, trans. and ed. Michael N. Forster (Cambridge: Cambridge University Press, 2002), 3-29.

102. Johann Gottfried Herder, "Gedichte von Sophie Mereau: Erstes Bändchen: Berlin 1800," in *Sämmtliche Werke: Zur Philosophie* und Geschichte (Stuttgart: Cotta, 1830), 21:392-98, here 394.

103. Herder, "Gedichte," 394.

104. Herder, "Gedichte," 393-94.

105. Herder, *Abhandlung*, 711; *Treatise*, 112.

106. Herder, *Abhandlung*, 769; *Treatise*, 127.

107. Herder, *Abhandlung*, 799; *Treatise*, 154.

108. Herder, *Abhandlung*, 723; *Treatise*, 88.

109. Herder, *Abhandlung*, 701; *Treatise*, 69.

110. Dorothea von Mücke, *Virtue and the Veil of Illusion: Generic Innovation and the Pedagogical Project in Eighteenth-Century Literature* (Stanford, CA: Stanford University Press, 1991), 167.

111. Herder, *Abhandlung*, 701; *Treatise*, 69.

112. Herder, *Abhandlung*, 707; *Treatise*, 74.

113. Herder, *Abhandlung*, 783; *Treatise*, 89.

114. See Sven Rausch, "Sophrosyne," in *Der Neue Pauly: Enzyklopädie der Antike*, ed. Hubert Cancik, Helmuth Schneider, and Manfred Landfester (first published online 2006); E. Heintel, "Besonnenheit," in *Historisches Wörterbuch der Philosophie*, vol. 1, ed. Joachim Ritter (Basel: Schwabe, 1971), 848-49.

115. Helen North, *Sophrosyne: Self-Knowledge and Self-Restraint in Greek Literature* (Ithaca, NY: Cornell University Press, 1966). North traces varying characterizations of sophrosyne, from soundness of mind to self-knowledge to prudence to moral virtue.

116. Heintel, "Besonnenheit," 849.

117. See Tanvi Solanki, "Aural Philology: Herder Hears Homer Singing," *Classical Receptions Journal* 12, no. 4 (October 2020): 401–24.

118. Herder, *Abhandlung*, 717; *Treatise*, 83.

119. Herder, *Abhandlung*, 719; *Treatise*, 85.

120. North, *Sophrosyne*, x.

121. Helen F. North, "The Mare, the Vixen, and the Bee: 'Sophrosyne' as the Virtue of Women in Antiquity," *Illinois Classical Studies* 2 (1977): 35–48, here 38.

122. Anne Carson, "The Gender of Sound," in *Glass, Irony, and God* (New York: New Directions, 1995), 119–39, here 126.

123. Herder, *Abhandlung*, 724; *Treatise*, 89.

124. von Mücke, *Virtue*, 169. Kittler also argues that in Herder, "the lamb stands for Woman." See Kittler, *Discourse Networks*, 39.

125. Herder, *Abhandlung*, 810; *Treatise*, 164.

126. Jürgen Trabant calls Herder's *Treatise* a "sensualist exegesis of the Bible," pointing out that the sheep scene is a human-centered version of Adam's naming of the animals. Jürgen Trabant, "Introduction: New Perspectives on an Old Academic Question," in *New Essays on the Origin of Language*, ed. Jürgen Trabant and Sean Ward (New York: Mouton de Gruyter, 2001), 1–17, here 2 and 5.

127. Fichte, "Ursprung," 97; "Linguistic Capacity," 119.

128. Johann Gottlieb Fichte, "Grundlage des Naturrechts nach Principien der Wissenschaftslehre," in *Werke*, ed. Reinhard Lauth and Hans Jacob, *Gesamtausgabe der Bayerischen Akademie der Wissenschaften* (Stuttgart: Friedrich Frommann, 1966), 3:291–460, here 3:348; Fichte, *Foundations of Natural Right*, ed. Frederick Neuhouser, trans. Michael Baur (Cambridge: Cambridge University Press, 2000), 38.

129. Bärbel Frischmann, "Fichte's Theory of Gender Relations in his *Foundations of Natural Right*," in *Rights, Bodies and Recognition: New Essays on Fichte's Foundations of Natural Right*, ed. Tom Rockmore and Daniel Breazeale (Burlington, VT: Ashgate, 2006), 152–65, here 156.

130. Adrian Daub, *Uncivil Unions: The Metaphysics of Marriage in German Idealism and Romanticism* (Chicago: University of Chicago Press, 2012), 53.

131. Johann Gottlieb Fichte, "Das System der Sittenlehre nach den Principien der Wissenschaftslehre," in *Werke*, ed. Reinhard Lauth and Hans Gliwitzky, *Gesamtausgabe der Bayerischen Akademie der Wissenschaften* (Stuttgart: Freidrich Frommann, 1977), 5:1–317, here 5:289; Fichte, *The System of Ethics: According to the Principles of the Wissenschaftslehre*, trans. and ed. Daniel Breazeale and Günter Zöller (Cambridge: Cambridge University Press, 2005), 312–13.

132. Fichte, "Sittenlehre," 288; *Ethics*, 312.

133. Fichte, "Sittenlehre," 288; *Ethics*, 312.

134. Johann Gottlieb Fichte, "Grundlage des Naturrechts nach Principien der Wissenschaftslehre: Zweyter Theil, oder Angewandtes Naturrecht," in *Werke*, ed. Reinhard Lauth and Hans Gliwitzky with Richard Schottky, *Gesamtausgabe der Bayerischen Akademie der Wissenschaften* (Stuttgart: Friedrich Fromm, 1970), 4:1–165, here 4:97; *Foundations*, 266.

135. Fichte, "Sittenlehre," 289; *Ethics*, 312.

136. Fichte, "Sittenlehre," 289; *Ethics*, 312.

137. Fichte, "Grundlage des Naturrechts," 4:113; *Foundations*, 282.

138. Fichte, "Sittenlehre," 289; *Ethics*, 313.

139. Fichte, "Sittenlehre," 290; *Ethics*, 314.

140. Müller-Sievers, *Self-Generation*, 87.

141. Fichte, "Grundlage des Naturrechts," 4:102; *Foundations*, 271.

142. Fichte, "Grundlage des Naturrechts," 4:102; *Foundations*, 271.

143. Fichte, "Grundlage des Naturrechts," 4:102; *Foundations*, 270.

144. Fichte, "Sittenlehre," 296; *Ethics*, 320.

145. Fichte, "Grundlage des Naturrechts," 4:99; *Foundations*, 268.

146. Fichte, "Sittenlehre," 291; *Ethics*, 315.

147. Fichte, "Sittenlehre," 291; *Ethics*, 315.

148. "In the union of the two sexes (and therefore, in the realization of the *whole* human being as a perfected product of nature), but also only in this union, is there to be found an external drive towards virtue." Fichte, "Grundlage des Naturrechts," 4:104; *Foundations*, 273.

149. Fichte, "Grundlage des Naturrechts," 102; *Foundations*, 271.

150. Stefani Engelstein, "The Allure of Wholeness: The Eighteenth-Century Organism and the Same-Sex Marriage Debate," *Critical Inquiry* 39, no. 4 (Summer 2013): 754–76, here 759.

151. Johann Gottlieb Fichte, "Einige Vorlesungen über die Bestimmung des Gelehrten," in *Werke*, 3:1–68, here 3:30; Fichte, "Some Lectures Concerning the Scholar's Vocation," in *Early Philosophical Writings*, trans. and ed. Daniel Breazeale (Ithaca, NY: Cornell University Press, 1988), 144–84, here 149.

152. Fichte, "Bestimmung des Gelehrten," 3:32; "Scholar's Vocation," 152.

153. Michelle Kosch, "Agency and Self-Sufficiency in Fichte's Ethics," *Philosophy and Phenomenological Research* 91, no. 2 (September 2015): 348–80, here 351.

154. Fichte, "Grundlage des Naturrechts," 4:100; *Foundations*, 269. My emphasis.

155. Fichte, "Grundlage des Naturrechts," 4:100; *Foundations*, 269.

156. Frischmann, "Gender Relations," 156.

157. Fichte, "Grundlage des Naturrechts," 4:135; *Foundations*, 304.

158. Fichte, "Grundlage des Naturrechts," 4:135; *Foundations*, 304.

159. Fichte, "Ursprung," 97; "Linguistic Capacity," 119.

160. Fichte, "Ursprung," 97; "Linguistic Capacity," 119.

161. Fichte, "Ursprung," 98; "Linguistic Capacity," 120.

162. Fichte, "Ursprung," 99; "Linguistic Capacity," 121.

163. Fichte, "Ursprung," 99; "Linguistic Capacity," 121.

164. Fichte, "Ursprung," 99; "Linguistic Capacity," 121.

165. Fichte, "Ursprung," 99; "Linguistic Capacity," 121.

166. Fichte, "Ursprung," 100; "Linguistic Capacity," 122.

167. Fichte, "Ursprung," 101; "Linguistic Capacity," 122.

168. Fichte, "Ursprung," 102; "Linguistic Capacity," 123.

169. Fichte, "Ursprung," 103; "Linguistic Capacity," 124.

170. Fichte, "Ursprung," 101–2; "Linguistic Capacity," 123.

171. Fichte, "Ursprung," 103; "Linguistic Capacity," 125.

172. Fichte, "Ursprung," 104; "Linguistic Capacity," 125.

173. Fichte, "Ursprung," 104; "Linguistic Capacity," 125.

174. Fichte, "Ursprung," 105; "Linguistic Capacity," 126.

175. Fichte, "Ursprung," 107; Fichte, "Linguistic Capacity," 127–28.

176. Fichte, "Linguistic Capacity," 129.

177. Fichte, "Ursprung," 109; "Linguistic Capacity," 129.

178. Fichte, "Ursprung," 107; "Linguistic Capacity," 127.

179. Andrew Fiala, "Fichte and the *Ursprache*," in *After Jena: New Essays on Fichte's Later Philosophy*, ed. Daniel Breazeale and Tom Rockmore (Evanston, IL: Northwestern University Press, 2008), 183–97, here 183.

180. Fiala, "Fichte and the *Ursprache*," 185.

181. Johann Gottlieb Fichte, *Reden an die deutsche Nation*, in *Werke*, ed. Reinhard Lauth, Erich Fuchs, Peter K. Schneider, Hans Georg von Manz, Ives Radrizzani, Martin Siegel, and Günter Zöller with Josef Beeler-Port, *Gesamtausgabe der Bayerischen Akademie der Wissenschaften* (Stuttgart: Friedrich Fromm, 2005), 10:1–298, here 144; Johann Gottlieb Fichte, *Addresses to the German Nation*, ed. Gregory Moore (Cambridge: Cambridge University Press, 2008), 48.

182. Fichte, *Reden*, 144; *Addresses*, 49.

183. Fichte, *Reden*, 150; *Addresses*, 53–54.

184. Étienne Balibar, "Fichte and the Internal Border: On *Addresses to the German Nation*," in *Masses, Classes, Ideas: Studies on Politics and Philosophy Before and After Marx* (New York: Routledge, 1994): 61–84, here 79.

185. Balibar, "Fichte," 81. Balibar continues his explanation: "The transcendental of language is thus not a *given* transcendental, in which thought is enclosed by categories or 'means of expression,' but a transcendental speech, which is at the same time the act of auto-constitution of thought."

186. Balibar, "Fichte," 79.

187. As Arash Abizadeh persuasively shows, Fichte's text is actually ambivalent about the extent to which people of other ethnicities can assimilate into German culture. While Fichte explicitly rejects "purity of descent" as constitutive of the nation, he often slides into the language of ethnic nationalism in his descriptions. Arash Abizadeh, "Was Fichte an Ethnic Nationalist? On Cultural Nationalism and Its Double," *History of Political Thought* 26, no. 2 (Summer 2005): 334–59.

188. Richard Wagner, "Das Judenthum in der Musik," in Frank Piontek, *Richard Wagners* "Das Judenthum in der Musik": *Text, Kommentar und Wirkungsgeschichte* (Beucha: Sax, 2017), 20–52, here 23–24; Wagner, "Judaism in Music," in *Judaism in Music and Other Essays*, trans. William Ashton Ellis (Lincoln: University of Nebraska Press, 1995), 75–122, here 84.

189. Daub, *Uncivil Unions*, 65.

190. Daub, *Uncivil Unions*, 66.

191. Wilhelm von Humboldt, "Ueber den Geschlechtsunterschied und dessen Einfluß auf die organische Natur [1794]," in *Gesammelte Schriften*, ed. Albert Leitzmann (Berlin: de Gruyter, 1968), 1:311–34, here 319.

192. Humboldt, "Geschlechtsunterschied," 282.

193. Wilhelm von Humboldt, "Ueber die männliche und weibliche Form [1795]," in *Gesammelte Schriften*, ed. Albert Leitzmann (Berlin: de Gruyter, 1968), 1:335–69, here 335.

194. Richter and Reill point to Humboldt's claim that perfection is possible only through the combination of masculine and feminine, and that neither gender could, as Reill puts it, "ever exist as an isolated thing in and for itself" (70). See Peter Hanns Reill, "The Scientific Construction of Gender and Generation in the German Late Enlightenment and in German Romantic *Naturphilosophie*," in *Reproduction, Race, and Gender in Philosophy and the Early Life Sciences*, ed. Susanne Lettow (Albany: State University of New York Press, 2014), 65–82; and Simon Richter, "Weimar Heteroclassicism: Wilhelm von Humboldt, Caroline von Wolzogen, and the Aesthetics of Gender," *Publications of the English Goethe Society* 81, no. 3 (October 2012): 137–51.

195. Humboldt, *Form*, 237.

196. Humboldt, "Geschlechtsunterschied," 289.

197. Humboldt, "Geschlechtsunterschied," 281.

198. Christina von Braun, "Männliche und weibliche Form in Natur und Kultur in der Wissenschaft," (presentation, "Humboldt Gespräche," Conference for the Bundeszentrale für politische Bildung, Berlin, Germany, June 7-8, 2007). Braun argues that Humboldt's construction of masculinity as (among other things) a "sparseness of matter" mirrors the construction of the nineteenth-century scientist, who is required "seinen Körper an der Garderobe [abzugeben], bevor er das Labor betritt" (1). See also Catriona MacLeod, *Embodying Ambiguity: Androgyny and Aesthetics from Winckelmann to Keller* (Detroit: Wayne State University Press, 1998).

199. Aristotle, *Generation of Animals*, trans. A. L. Peck, Loeb Classical Library 366 (Cambridge, MA: Harvard University Press, 1942), 729 a, p. 111.

200. Aristotle, *Generation of Animals*, 729 b, p. 113.

201. MacLeod, *Embodying Ambiguity*, 51.

202. See MacLeod, *Embodying Ambiguity*, 49.

203. Humboldt, *Form*, 259-60.

204. MacLeod, *Embodying Ambiguity*, 51.

205. Humboldt, *Form*, 260.

206. Humboldt, "Geschlechtsunterschied," 291.

207. Wilhelm von Humboldt, "Plan einer vergleichenden Anthropologie," in *Gesammelte Schriften*, ed. Albert Leitzmann (Berlin: de Gruyter, 1986), 1:377-410, here 406. As quoted in Claudia Honegger, *Die Ordnung der Geschlechter: Die Wissenschaften vom Menschen und das Weib, 1750-1850* (Frankfurt a. M.: Campus, 1991), 185.

208. For more on the history of Humboldt's Kawi manuscript, see Wilhelm von Humboldt, "Note on the Text" in *On Language: On the Diversity of Human Language Construction and Its Influence on the Mental Development of the Human Species*, trans. Peter Heath, ed. Michael Losonsky (Cambridge: Cambridge University Press, 1999), xl-xli.

209. Humboldt, *Verschiedenheiten*, 434-35; *Diversity*, 60.

210. Humboldt, *Verschiedenheiten*, 448; *Diversity*, 70.

211. Humboldt, *Verschiedenheiten*, 429; *Diversity*, 56.

212. Humboldt, *Verschiedenheiten*, 548; *Diversity*, 143.

213. Peter Hanns Reill, "Science and the Construction of the Cultural Sciences in Late Enlightenment Germany: The Case of Wilhelm von Humboldt," *History and Theory* 33, no. 3 (October 1994): 345-66, here 357.

214. Reill, "Science and the Construction of the Cultural Sciences," 358. Reill elaborates on this in a different essay, writing of Humboldt's attraction to "strong females," such as a "ferry woman who awakened in him powerful sado-masochistic fantasies" ("Scientific Construction of Gender," 71).

215. Jürgen Trabant, "Nachwort," in Wilhelm von Humboldt, *Über die Sprache* (Tübingen: A. Francke, 1994), 201-17, here 207. See also Jürgen Trabant, *Apeliotes, oder, Der Sinn der Sprache: Wilhelm von Humboldts Sprach-Bild* (Munich: W. Fink, 1986), 21.

216. Trabant, *Apeliotes*, 18.

217. Trabant, *Apeliotes*, 16-17.

218. James W. Underhill, *Humboldt, Worldview and Language* (Edinburgh: Edinburgh University Press, 2009), 81.

219. Albisetti, *Schooling German Girls and Women*, 15.

220. Kittler, *Discourse Networks*, 25.

221. Humboldt, *Verschiedenheiten*, 418; *Diversity*, 49.

222. Humboldt, *Verschiedenheiten*, 416; *Diversity*, 48.

223. Humboldt, *Verschiedenheiten*, 418; *Diversity*, 49.

224. Hans Aarsleff, "The Context and Sense of Humboldt's Statement that Language 'Ist kein Werk (Ergon), sondern eine Tätigkeit (Energeia),'" in *History of Linguistics 2002: Selected Papers from the Ninth International Conference on the History of the Language Sciences, 27-30 August 2002, São Paulo—Campinas*, ed. Eduardo Guimarães and Diana Luz Pessoa de Barros (Amsterdam: John Benjamins, 2007), 197–206, here 198.

225. See Kristina Mendicino, *Prophecies of Language: The Confusion of Tongues in German Romanticism* (New York: Fordham University Press, 2016), 75–76.

226. Humboldt, *Verschiedenheiten*, 413; *Diversity*, 44. To put it more simply, language as *Energeia* is "the transformation of matter into idea and of articulated sound into thought" (James Stam's summary). James H. Stam, *Inquiries into the Origin of Language: The Fate of a Question* (New York: Harper & Row, 1976), 210.

227. Humboldt, *Verschiedenheiten*, 421; *Diversity*, 51.

228. Humboldt, *Verschiedenheiten*, 414; *Diversity*, 46, translation modified.

229. Humboldt, *Verschiedenheiten*, 415; *Diversity*, 47.

230. Humboldt, *Verschiedenheiten*, 415; *Diversity*, 46.

231. Humboldt, *Verschiedenheiten*, 421; *Diversity*, 46.

232. Humboldt, *Verschiedenheiten*, 422; *Diversity*, 51–52.

233. Humboldt, *Verschiedenheiten*, 421; *Diversity*, 52.

234. Humboldt, *Verschiedenheiten*, 426; *Diversity*, 55.

235. Humboldt, *Verschiedenheiten*, 421; *Diversity*, 51.

236. Humboldt, *Verschiedenheiten*, 421; *Diversity*, 51.

237. Humboldt, *Verschiedenheiten*, 435; *Diversity*, 60–61.

238. Humboldt, *Verschiedenheiten*, 435–36; *Diversity*, 61.

239. Lia Formigari, "Idealism and Idealistic Trends in Linguistics and in the Philosophy of Language," in *Der epistemologische Kontext neuzeitlicher Sprach- und Grammatiktheorien*, vol. 1 of *Sprachtheorien der Neuzeit*, ed. Peter Schmitter (Tübingen: Gunter Narr, 1999), 230–53, here 234.

240. Formigari, "Idealism and Idealistic Trends," 235.

241. Rousseau, *Émile*, 490; *Emile*, 376.

242. Joan Wallach Scott, *Only Paradoxes to Offer: French Feminists and the Rights of Man* (Cambridge, MA: Harvard University Press, 1996), 19.

243. Friedrich Schlegel, *Kritische Friedrich-Schlegel-Ausgabe* 8:169; *On the Language and Philosophy of the Indians*, in *The Aesthetic and Miscellaneous Works*, trans. E. J. Millington (London: Henry G. Bohn, 1849), 424–626, here 454. Translation modified.

244. Schlegel, *Kritische Friedrich-Schlegel-Ausgabe* 8:169; *Language and Philosophy of the Indians*, 454.

245. In contrast to someone like Humboldt, Schlegel has been received as a more progressive theorist of gender difference. But especially later in his career, he proposed a rigid and biologically grounded conception of the sexes not so different from that of Humboldt and others like him, as Catriona MacLeod shows. See MacLeod, *Embodying Ambiguity*, 66–90, and Sarah Friedrichsmeyer, *The Androgyne in Early German Romanticism: Friedrich Schlegel, Novalis and the Metaphysics of Love* (Bern: Peter Lang, 1983), 128–29.

246. Grimm, *Ursprung der Sprache*, 37.

247. Grimm alludes to Herder in the last paragraph of his lecture. It was also common for linguistic texts to take a quote from Herder as an epigraph, as occurs, for instance, in Johann Karl Friedrich Rinne's *Die natürliche Entstehung der Sprache aus dem Gesichtspuncte der historischen oder vergleichenden Sprachwissenschaft* (Erfurt: Friedrich Wilhelm Otto, 1834). For more on scholars who were influenced by and even plagiarized Herder, see Stam, *Inquiries*, 166.

248. On the different iterations of Herder's philosophy of language, see Michael N. Forster, "Gods, Animals, and Artists: Some Problem Cases in Herder's Philosophy of Language," in *After Herder: Philosophy of Language in the German Tradition* (Oxford: Oxford University Press, 2010), 91–130.

249. The case of the linguist August Schleicher is interesting in this regard. When Schleicher ridicules an unnamed "Danish colleague" in 1860 for asserting that "the inventor of language was a man, not a woman!" this is not because Schleicher champions women's abilities, but because he objects to the idea that we can talk about *one* inventor of language at all. Schleicher espouses a polygenetic theory of language origin. Schleicher, *Die deutsche Sprache* (Stuttgart: Cotta, 1860), 39.

250. Heymann Steinthal, *Abriss der Sprachwissenschaft: Einleitung in die Psychologie und Sprachwissenschaft* (Berlin: Ferd. Dümmler, 1871), 89–90.

251. Steinthal, *Abriss*, 84.

252. Steinthal, *Abriss*, 84.

253. Humboldt, *Verschiedenheiten*, 421; *Diversity*, 51.

254. Johann Gottfried Herder, "Paramythien: Dichtungen aus der griechischen Sage," in *Werke in zehn Bänden*, ed. Ulrich Gaier (Frankfurt a.M.: Deutscher Klassiker, 1990), 3:697–724, here 708; "Paramythia: From the German of Herder," *Belfast Monthly Magazine* 2, no. 9 (April 30, 1809): 262–66, here 264.

255. Herder, "Paramythien," 709; "Paramythia," 264.

256. Herder, "Paramythien," 709; "Paramythia," 264.

Chapter Two

1. Wilhelm von Humboldt, "Ueber die Verschiedenheiten des menschlichen Sprachbaues (1827–1829)," in *Werke in fünf Bänden*, ed. Andreas Flitner and Klaus Giel (Darmstadt: Wissenschaftliche Buchgesellschaft, 2002), 3:144–367, here 253.

2. Humboldt, "Verschiedenheiten," 253.

3. See, for instance, Charles de Rochefort, *Histoire naturelle et morale des îles Antilles de l'Amérique: Enrichie de plusieurs belles figures des raretez les plus considerables qui y sont d'écrites, avec un vocabulaire Caraïbe* (Rotterdam: Arnould Leers, 1658), 395. Rochefort reports that this explanation comes from the Island Caribs themselves. A similar distinction between multiple languages is reported by Mathias DuPuis in *Relation de l'establissement d'une colonie françoise dans la Gardeloupe, isle de l'Amérique, et des moeurs des sauvages* (1652; Basse-Terre: Société d'Histoire de la Guadeloupe, 1972), 196.

4. Neil L. Whitehead, "Introduction: The Island Carib as Anthropological Icon," in *Wolves from the Sea: Readings in the Anthropology of the Native Caribbean*, ed. Neil L. Whitehead (Leiden: KITLV Press, 1995), 9–22, here 11.

5. Peter Hulme and Neil L. Whitehead, "The Letter of Columbus (1493)," in *Wild Majesty: Encounters with Caribs from Columbus to the Present Day: An Anthology*, ed. Peter Hulme and Neil L. Whitehead (Oxford: Clarendon Press, 1992), 9–16, here 9.

6. Whitehead, "Introduction: The Island Carib as Anthropological Icon," 10.

7. Missionaries believed that the people native to the Lesser Antilles were descendants of mainland Caribs (known today as Karìna or Kalina). This is a point now debated in modern scholarship. Several critics have argued that the "Island Caribs" were actually Arawaks, the "Carib" language spoken on the islands actually an Arawakan language, and that differences between Arawaks and Caribs were exaggerated by early European settlers in order to validate their takeover of the islands. For the purposes of my book, however, to avoid confusion, I will maintain the name *Carib* or *Island Carib*, since these are the terms used by my primary sources. For more on the debate about the "Island Carib" language, see the sources mentioned in the following notes as well as the work of Douglas Taylor: "Languages and Ghost-Languages of the West-Indies," *International Journal of American Linguistics* 22, no. 2 (April 1956): 180–83; and, with Berend J. Hoff, "The Linguistic Repertory of the Island-Carib in the Seventeenth Century: The Men's Language: A Carib Pidgin?" *International Journal of American Linguistics* 46, no. 4 (October 1980): 301–12. See also Berend J. Hoff, "Language Contact, War, and Amerindian Historical Tradition: The Special Case of the Island Carib," in *Wolves from the Sea*, ed. Whitehead, 27–60.

8. Douglas Taylor and Berend J. Hoff, for instance, argue that the "men's" language of the Caribs was actually a pidgin used for trading. Vincent O. Cooper contends that "Island Carib" should not be considered a "homogeneous language or a pidgin but a dynamic communication system that reflected the sociocultural flexibility needed for adaptation to maritime travel, trade, warfare, and natural disaster." Similarly, Neil Whitehead writes that "Island Arawak" and "Island Carib" were not "opposed or exclusive populations but aspects of a Caribbean and South American matrix in which a plurality of identities were represented at different times." See Vincent O. Cooper, "Language and Gender among the Kalinago of Fifteenth-century St. Croix," in *The Indigenous People of the Caribbean*, ed. Samuel M. Wilson (Gainesville: University Press of Florida, 1997), 186–96, here 194; Louis Allaire's chapter in the same volume, "The Caribs of the Lesser Antilles," 177–85; and Whitehead, "Carib as Icon," 12.

9. Whitehead, "Carib as Icon," 10. Rousseau, for instance, refers to Caribs in his characterization of "the savage" in the *Discourse on Inequality*: "His soul, which nothing disturbs, gives itself up entirely to the consciousness of its actual existence [. . .] Such is, even at present, the degree of foresight in the Caribbean: he sells his cotton bed in the morning, and comes back in the evening, with tears in his eyes, to buy it back, not having foreseen that he should want it again the next night." Jean-Jacques Rousseau, *Discourse on Inequality: On the Origin and Basis of Inequality among Men* (Auckland: Floating Press, 2009), 30.

10. See Whitehead, "Carib as Icon," and Stephan Lenik, "Carib as a Colonial Category: Comparing Ethnohistoric and Archaeological Evidence from Dominica, West Indies," *Ethnohistory* 59, no. 1 (Winter 2012): 79–107.

11. DuPuis, *Relation*, 196.

12. DuPuis, *Relation*, 196.

13. DuPuis, *Relation*, 202.

14. DuPuis, *Relation*, 203.

15. From Gaetano DeLeonibus, "Raymond Breton's 1665 'Dictionaire caraïbe-françois,'" *French Review* 80, no. 5 (April 2007): 1044–55, here 1045. DeLeonibus provides extensive background information on Breton's time in the West Indies and the contemporaneous context of lexicography and "of the aesthetic and ideological discourses that usurped the cultural arena under French neo-classicism" (1045).

16. Raymond Breton, *Dictionnaire françois-caraïbe* (Auxerre: Gilles Bouquet, 1666), 5. Breton's Carib-French dictionary (*Dictionnaire caraïbe-françois*) was published one year earlier, in 1665, but it is only in the French-Carib dictionary that he uses the term *langage des femmes*.

17. Unlike Breton, neither Rochefort nor Du Tertre wrote a personal account of their time in the Caribbean and consequently less is known about their background. Rochefort, a French-speaking pastor of the Walloon church in Holland, was first sent to the Antilles in 1636 and lived there for twelve years, until 1648. Du Tertre, a French Dominican friar, made three missionary trips to the Antilles between 1640 and 1657. Both of their texts appear to have met success upon publication, as multiple editions were reissued in the second half of the seventeenth century. Rochefort's text was also quickly translated into Dutch, German, and English. See Benoît Roux, "Le pasteur Charles de Rochefort et l'*Histoire naturelle et morale des îles Antilles de l'Amérique*," *Cahiers d'Histoire de l'Amérique Coloniale* 5 (2011): 175–216. For more on both Rochefort and Du Tertre, see the introductions to the excerpts from their texts in Whitehead, *Wild Majesty*.

18. For a history of the Spanish portrayal of Amerindians, see John F. Moffitt and Santiago Sebastián, *O Brave New People: The European Invention of the American Indian* (Albuquerque: University of New Mexico Press, 1998) as well as the Spanish sources in Whitehead, *Wild Majesty*.

19. Jean-Baptiste Du Tertre, *Histoire générale des Antilles habitées par les François* (Paris: Thomas Iolly, 1667), 2:357.

20. Philip P. Boucher, *Cannibal Encounters: Europeans and Island Caribs, 1492–1763* (Baltimore: Johns Hopkins University Press, 1992), 55.

21. Du Tertre, *Histoire générale* 2:2, my emphasis.

22. Du Tertre, *Histoire générale*, 2:361–62.

23. Du Tertre, *Histoire générale*, 2:361.

24. Rochefort, *Histoire naturelle et morale*, 394–95.

25. Rochefort, *Histoire naturelle et morale*, 392.

26. Rochefort, *Histoire naturelle et morale*, 393.

27. Rochefort, *Histoire naturelle et morale*, 399.

28. Rochefort, *Histoire naturelle et morale*, 397.

29. Rochefort, "Epistre a Messire Iaqves Amproux, Seigneur de Lorme, Conseiller du Roy en ses Conseils d'Etat & Privé, & Intendant de ses Finances," in *Histoire naturelle et morale* (no pagination).

30. Mary Baine Campbell, *Wonder and Science: Imagining Worlds in Early Modern Europe* (Ithaca, NY: Cornell University Press, 1999), 289.

31. Joseph-François Lafitau, *Moeurs des sauvages amériquains comparées aux moeurs des premiers temps* (Paris: Saugrain l'aîné, Charles-Estienne Hochereau, 1724), 1:50–51.

32. There is a second section in Herodotus's *Histories* where the notion of men's and women's languages is at issue, which Lafitau does not mention—the story of the Amazons and the Scythians. According to Herodotus, their intermarriage originally resulted in a linguistic division, but the Amazons quickly adapted to the language of the Scythians, forsaking their "women's language": "The two sides joined forces and lived together, forming couples consisting of a Scythian man and the Amazon with whom he had first had sex. The men found the women's language impossible to learn, but the women managed the men's language." Herodotus, *The Histories*, trans. Robin Waterfield (Oxford: Oxford University Press, 1998), 272.

33. For more on Lafitau's writing about language, see Sylviane Albertan-Coppola, "Les langues dans les *Moeurs des sauvages* de Lafitau," *Dix-Huitième Siècle* 22 (1990): 127–38.

34. Ovid, *Metamorphoses: A New Verse Translation*, trans. David Raeburn (London: Penguin, 2004), 237. Ovid's description of the rape of Philomela is just one of many accounts that circulated in the ancient Greek and Roman worlds. For the history of this myth, see the entry for "Tereus" in the *Encyclopedia of Greek and Roman Mythology*, 460.

35. Campbell, *Wonder and Science*, 289.

36. George R. Healy, "The French Jesuits and the Idea of the Noble Savage," *William and Mary Quarterly* 15, no. 2 (April 1958): 143–67, here 161.

37. Sara Petrella, "Femmes à poils: Réception et actualisation d'un cliché dans les *Moeurs des sauvages ameriquains* de Joseph-François Lafitau," in *La Plume et le calumet: Joseph-François Lafitau et les "sauvages ameriquains,"* ed. Mélanie Lozat and Sara Petrella (Paris: Garnier, 2019), 139–51, here 150.

38. William A. Pettigrew, *Freedom's Debt: The Royal African Company and the Politics of the Atlantic Slave Trade, 1672–1752* (Chapel Hill: Omohundro Institute and University of North Carolina Press, 2013).

39. Francis Moore, *Travels into the Inland Parts of Africa: Containing a Description of the Several Nations for the Space of Six Hundred Miles Up the River Gambia* [. . .] (London: Edward Cave, 1738), 28.

40. The term *cultural racist* in reference to Meiners comes from Marilyn Katz. Katz distinguishes between "cultural racism" and "scientific racism," which is "a theory about the existence of biologically distinct forms of humanity, distinguished from one another not on the basis of geography, climate, customs, or manners, which are susceptible to change and influence, but on the basis of an inherent and immutable biological essence." Marilyn Katz, "Ideology and the 'Status of Women' in Ancient Greece," *History and Theory* 31, no. 4 (December 1992): 70–97, here 91.

41. Franklin makes this claim about both Meiners and Joseph Alexandre Pierre Ségor, who wrote a similar book about the history of women in French. Ségor, however, does not make mention of men's and women's languages. See Caroline Franklin, "'Quiet Cruising o'er the Ocean Woman': Byron's 'Don Juan' and the Woman Question," *Studies in Romanticism* 29, no. 4 (Winter 1990): 603–31, here 606.

42. Christoph Meiners, *Geschichte des weiblichen Geschlechts* (Hannover: Helwingsche Hofbuchhandlung, 1788): 1:v.

43. Meiners, *Geschichte des weiblichen Geschlechts*, 1:70–71, my emphasis.

44. Alexander von Humboldt, *Relation historique du Voyage aux Régions équinoxiales du Nouveau Continent: Fait en 1799, 1800, 1801, 1802, 1803 et 1804 par Al. de Humboldt et A. Bonpland, rédigé par Alexandre de Humboldt* (1814–31; Stuttgart: F. A. Brockhaus, 1970), 3:10.

45. Humboldt, *Relation historique*, 3:10–11. The exact quote from Cicero, which Humboldt footnotes, is from Book III of *De Oratore*. The English translation reads: "For my part, when I hear my mother-in-law Laelia, (for it is easier for women to keep the purity of antiquity, because, by keeping less company than men, they always stick to what they first learned,) I think that I am conversing with Plautus or Naevius [. . .]." Cicero, *De Oratore*, trans. William Guthrie (Oxford: Henry Slatter, 1840), 225.

46. Humboldt, *Relation historique* 3:11, my emphasis.

47. Oliver Lubrich, "Alexander von Humboldt: Revolutionizing Travel Literature," *Monatshefte* 96, no. 3 (Fall 2004): 360–87. For Lubrich, Humboldt's refusal to find a "master signifier" for the geographical area he describes renders his text neither "occidentalist" nor "latinoamericanist" (369).

48. Aristotle, *History of Animals: Volume 3, Books 7-10*, ed. and trans. D. M. Balme, Loeb Classical Library 439 (Cambridge, MA: Harvard University Press, 1991), 608 b, lines 1-14.

49. Aristotle, *Minor Works*, trans. W. S. Hett, Loeb Classical Library 307 (Cambridge, MA: Harvard University Press, 1936), 95. "Those who have a deep, braying voice are insolent; witness the asses. But those whose voice begins deep and ends on a high-pitched note are despondent and plaintive; this applies to cattle and is similar to the voice. But those who talk with high-pitched, gentle and broken voices are morbid; this applies to women and is appropriate. Those who have a loud, deep voice, with a clear note, may perhaps be compared to brave dogs and are in conformity with their nature [. . .]" (133).

50. Aristotle, *Problems*, trans. E. S. Forster, in *Complete Works of Aristotle: The Revised Oxford Translation*, vol. 2, ed. Jonathan Barnes (Princeton, NJ: Princeton University Press, 1984), 900 b. The text explains further that the air set in motion by these persons "has only one dimension" and "is very small in quantity," which makes the air thin and produces a shrill voice.

51. Thorsten Fögen, "Gender-Specific Communication in Graeco-Roman Antiquity," *Historiographia Linguistica* 32, no. 2-3 (2004): 199-276, here 219. For more on gendered conceptions of speech in antiquity, see Adam Knowles, "A Genealogy of Silence: *Chōra* and the Placelessness of Greek Women," *philoSOPHIA* 5, no. 1 (Winter 2015): 1-24. See also André Lardinois and Laura McClure, eds., *Making Silence Speak: Women's Voices in Greek Literature and Society* (Princeton, NJ: Princeton University Press, 2001).

52. Carson, "The Gender of Sound," 119-39.

53. Plato, *Cratylus*, trans. C. D. C. Reeve, in *Complete Works*, ed. John M. Cooper (Indianapolis: Hackett, 1997), 418b-c.

54. Cicero, *De Oratore*, 225.

55. Semonides, "Fragments," in *Greek Iambic Poetry: From the Seventh to the Fifth Centuries BC*, ed. and trans. Douglas E. Gerber, Loeb Classical Library 259 (Cambridge, MA: Harvard University Press, 1999), 298-341, here 305-7.

56. Semonides, "Fragments," 311.

57. Juvenal, *The Satires of Juvenal*, with introduction and notes by A. F. Cole (London: J. M. Dent, 1906), 125.

58. See Sandy Bardsley, *Venomous Tongues: Speech and Gender in Late Medieval England* (Philadelphia: University of Pennsylvania Press, 2006).

59. See Kirilka Stavreva, *Words Like Daggers: Violent Female Speech in Early Modern England* (Lincoln: University of Nebraska Press, 2015).

60. Emily Butterworth, *The Unbridled Tongue: Babble and Gossip in Renaissance France* (Oxford: Oxford University Press, 2016), 173.

61. See Barbara Krug-Richter, "'Weibergeschwätz'? Zur Geschlechtsspezifik des Geredes in der Frühen Neuzeit," in *Weibliche Rede—Rhetorik der Weiblichkeit: Studien zum Verhältnis von Rhetorik und Geschlechterdifferenz*, ed. Doerte Bischoff and Martina Wagner-Egelhaaf (Freiburg i.Br.: Rombach, 2003), 301 and 319.

62. The Christian argument for women's silence was typically grounded in two key passages, I Corinthians 14:34-35 and I Timothy 2:12-14. I Corinthians asserts that "Women should be silent in the churches. For they are not permitted to speak, but should be subordinate, as the law also says. If there is anything they desire to know, let them ask their husbands at home. For it is shameful for a woman to speak in the church," while I Timothy maintains, "I permit no woman to teach or have authority over a man; she is to keep silent. For Adam was formed first, then Eve; and Adam was not deceived, but the woman was deceived and became a transgressor."

The New Oxford Annotated Bible, ed. Michael D. Coogan (Oxford: Oxford University Press, 2010), 2019 and 2087. Martin Luther also emphasized the importance of women's silence in public spaces. As recounted in his *Tischreden*, Luther asserts that women are by nature overly talkative, and that they should restrict their volubility to inside the household. For more on Luther and female speech, see Birgit Althans, *Der Klatsch, die Frauen und das Sprechen bei der Arbeit* (Frankfurt a.M.: Campus, 2000), 33–34.

63. Thomas Murner, "Das klapper benckly," in *Schriften*, ed. Franz Schultz (Berlin: de Gruyter, 1925), 3:88–89, here 89. Quoted in Pia Holenstein and Norbert Schindler, "Geschwätzgeschichte(n): Ein kulturhistorisches Plädoyer für die Rehabilitierung der unkontrollierten Rede," in *Dynamik der Tradition*, ed. Richard van Dülmen (Frankfurt a.M.: Fischer, 1992), 41–108, here 57. On the places associated with *Weibergeschwätz*, see also Althans, *Der Klatsch*, chapters 1 and 2; and Krug-Richter.

64. Holenstein and Schindler, "Geschwätzgeschichte(n)," 47–50. Their source for the aphorisms is Karl Simrock, *Die Deutschen Sprichwörter gesammelt* (Frankfurt a.M.: H. L. Brönner, 1846).

65. In facing the ire of German language purists, however, women are in good company. Other groups blamed for the degeneration of German include merchants, soldiers, lawyers, astrologers, preachers, chimney sweeps, ratcatchers, aristocrats, and peasants. See William Jervis Jones, *Images of Language: Six Essays on German Attitudes to European Languages from 1500 to 1800* (Amsterdam: John Benjamins, 1999), 37–38.

66. Jones, *Images of Language*, 36.

67. Jones, *Images of Language*, 75–76.

68. Robert Muchembled, *Smells: A Cultural History of Odours in Early Modern Times*, trans. Susan Pickford (Cambridge: Polity Press, 2020), 68.

69. As cited in Muchembled, *Smells*, 68.

70. Jacques Olivier [Alexis Trousset], *Alphabet de l'imperfection et malice des femmes* (Paris: Jean Petit-Pas, 1617), 93. Quoted and translated by Butterworth in *The Unbridled Tongue*, 174.

71. Patricia Howell Michaelson, *Speaking Volumes: Women, Reading, and Speech in the Age of Austen* (Stanford, CA: Stanford University Press, 2002), 25.

72. Michaelson, *Speaking Volumes*, 26.

73. Giambattista Vico, *The New Science of Giambattista Vico*, trans. Thomas Goddard Bergin and Max Harold Fisch (Ithaca, NY: Cornell University Press, 1984), 457.

74. Joachim Heinrich Campe, *Väterlicher Rath für meine Tochter* (Frankfurt a.M.: Cotta, 1789), 450–51.

75. Florence Hartley, *The Ladies' Book of Etiquette and Manual of Politeness: A Complete Hand Book for the Use of the Lady in Polite Society* (Boston: J. S. Locke, 1876), 17.

76. *Das häusliche Glück: Vollständiger Haushaltungsunterricht nebst Anleitung zum Kochen für Arbeiterfrauen* (M. Gladbach: A. Riffarth, 1905), 18–19. Reproduced in Holenstein and Schindler, "Geschwätzgeschichte(n)," 106–7.

77. Patricia Parker, *Literary Fat Ladies: Rhetoric, Gender, Property* (London: Methuen, 1987), 104.

78. Tonger-Erk, *Actio*, 313. Moreover, Tonger-Erk shows that, in the ancient world, oratory was restricted to men with no counterconcept of female eloquence. See *Actio*, 61.

79. Ruth B. Bottigheimer, *Grimm's Bad Girls and Bold Boys: The Moral and Social Vision of the Tales* (New Haven, CT: Yale University Press, 1987), 53.

80. See the section on the Amazons and the Scythians in *Les neuf livres des histoires d'Hérodote, prince et premier des Historiographes Grecz, intitulez du nom des Muses*, trans. Pierre Saliat (Paris: Jean de Roigny, 1556), 108.

81. The 1690 *Dictionaire universel* by Antoine Furetière lists "voix feminine" as an example under the entry "Feminin": "Les châtrez ont la voix *feminine*" (the castrated have a feminine voice; vol. 2, no pagination). The text makes no mention of the *langage des femmes* or any similar term. There is no entry for *voix féminine, langage des femmes,* or anything about women under the entries "Langue" and "Langage" in Jean Nicot's *Le Thresor de la langue francoyse* (1606) or the *Dictionnaire de l'Académie françoise* (first edition 1694). The term *langage des hommes* appears often in texts from this period, but this is *hommes* as in *mankind.*

82. There is no entry for *women's language* or any similar terms in Nathan Bailey's 1756 *Universal Etymological English Dictionary* or Oxford English Dictionary. There is likewise no entry in Samuel Johnson's *Dictionary of the English Language* (first published 1755). Johnson does, however, provide alternative definitions for several words in "women's cant."

83. Josua Maaler, *Die Teütsch spraach. Alle wuerter/ namen/ uñ arten zů reden in Hochteütscher spraach* (Tiguri: Christoph Froschoverus, 1561). Maaler's book is the first dictionary to "consist solely of an alphabetical German-Latin word list," according to William Jervis Jones. See Jones, "Lingua teutonum victrix? Landmarks in German Lexicography (1500–1700)," *Histoire Epistémologie Langage* 13, no. 2 (1991): 131–52.

84. Kaspar von Stieler, *Der teutschen Sprache Stammbaum und Fortwachs, oder Teutscher Sprachschatz* (Nürnberg: Johann Hofmann, 1691), 2102. Stieler's dictionary was unprecedented for its time, containing more words than any German dictionary compiled previously.

85. The exact definition reads: "weibersprache, *f. verächtlich im täglichen leben*: jungfer- und weibersprache, *sermo tinniens, vox acuta* STIELER 2102, *objectiv in wiss. würdigung*: Caraiben, ausgestorben. Weibersprache am ausgedehntesten w. v. HUMBOLDT 5 (1823) 26." *Deutsches Wörterbuch von Jacob und Wilhelm Grimm*, 16 vols. (Leipzig: S. Hirzel, 1854–1960), https://www .dwds.de/wb/dwb/weibersprache.

86. Stieler, *Der teutschen Sprache*, 2102. This definition is also reproduced in the Grimms' *Wörterbuch.*

87. Stieler, *Der teutschen Sprache*, 2167.

88. According to Benoît Roux, "Charles de Rochefort." This early German translation furthermore does not use the term *Weibersprache* to refer to the women's language. Charles de Rochefort, *Historische Beschreibung der Antillen Inseln in America gelegen: In sich begreiffend deroselben Gelegenheit, darinnen befindlichen natürlichen Sachen, sampt deren Einwohner Sitten und Gebräuchen* (Frankfurt a.M.: Wilhelm Serlins, 1668), 2:296. I thank the John J. Burns Library of Rare Books and Special Collections at Boston College for providing access to the German translation of Rochefort's text.

89. On the importance of empiricism and exactitude for nineteenth-century language science, see Andreas Gardt, *Geschichte der Sprachwissenschaft in Deutschland* (New York: de Gruyter, 1999), 268–71, and E. F. K. Koerner, "The Natural Science Impact on Theory Formation in 19th and 20th Century Linguistics," in *Professing Linguistic Historiography* (Philadelphia: John Benjamins, 1995), 47–76.

90. Herder, *Abhandlung,* 757; *Treatise,* 117.

91. Herder, *Abhandlung,* 756; *Treatise,* 117.

92. Herder, *Abhandlung,* 791; *Treatise,* 147, translation modified.

93. Wilhelm von Humboldt, "Ueber das vergleichende Sprachstudium in Beziehung auf die verschiedenen Epochen der Sprachentwicklung," in *Werke,* ed. Andreas Flitner and Klaus Giel (Darmstadt: Wissenschaftliche Buchgesellschaft, 2002), 3:1–24, here 20.

94. Humboldt, "Sprachstudium," 1.

95. Humboldt, "Sprachstudium," 6.

96. The German terms are "Data" (Humboldt, "Sprachstudium," 8) and "strengere faktische Prüfung" (Humboldt, "Sprachstudium," 9).

97. Humboldt, "Sprachstudium," 6.

98. Humboldt, "Sprachstudium," 8.

99. Humboldt, "Sprachstudium," 8.

100. Wilhelm von Humboldt, *Prüfung der Untersuchungen über die Urbewohner Hispaniens vermittelst der vaskischen Sprache* (Berlin: Ferdinand Dümmler, 1821).

101. Humboldt, "Sprachstudium," 11.

102. Georg Friedrich Benecke, review of *Deutsche Grammatik* by Jacob Grimm, *Göttingische gelehrte Anzeigen* 201 (December 19, 1822): 2001–8, here 2002–3. Quoted in Koerner, *Professing*, 52.

103. See Humboldt, "Sprachstudium," 1. See also Reill, "Science and the Construction of the Cultural Sciences"; and Irmline Veit-Brause, "Scientists and the Cultural Politics of Academic Disciplines in Late 19th-Century Germany: Emil Du Bois-Reymond and the Controversy over the Role of the Cultural Sciences," *History of the Human Sciences* 14, no. 4 (November 2001): 31–56.

104. Pourciau, *The Writing of Spirit*, 44.

105. Humboldt, "Verschiedenheiten," 253.

106. Humboldt, "Verschiedenheiten," 253.

107. Rousseau, *Émile*, 470; *Emile*, 361.

108. Humboldt, "Verschiedenheiten," 254.

109. Catherine Hobbs Peaden, "Condillac and the History of Rhetoric," *Rhetorica: A Journal of the History of Rhetoric* 11, no. 2 (Spring 1993): 135–56, here 140.

110. Falko Schnicke, *Die männliche Disziplin: Zur Vergeschlechtlichung der deutschen Geschichtswissenschaft 1780–1900* (Göttingen: Wallstein, 2015), 93. See also Bonnie G. Smith, *The Gender of History: Men, Women, and Historical Practice* (Cambridge, MA: Harvard University Press, 2000).

111. Paul Robinson Sweet, *Wilhelm von Humboldt: A Biography* (Columbus: Ohio State University Press, 1980), 1:165–70.

112. Quoted in Sweet, *Humboldt*, 1:170.

113. Quoted in Sweet, *Humboldt*, 1:169.

114. Humboldt, "Verschiedenheiten," 255.

115. Humboldt, "Verschiedenheiten," 255.

116. Humboldt, "Verschiedenheiten," 225.

117. Humboldt, "Verschiedenheiten," 254–55, my emphasis.

118. The first English edition of *Das Weib* appeared in 1935, issued by the publisher William Heinemann in London.

119. The most significant difference between Ploss's first edition and those that follow is the addition of illustrations. In 1887, two years after Ploss's death, Ploss's colleague Max Bartels edits a new edition that includes pictures. His son, Paul Bartels, then takes over the editing; later, Ferdinand von Reitzenstein adds images in the last German edition, expanded to three volumes, in 1927. For more on the publication history, see Paula Weideger's introduction to her 1985 English translation, *History's Mistress: A New Interpretation of a Nineteenth-Century Ethnographic Classic* (New York: Viking Penguin, 1985). I am working here with the 1908 edition: Heinrich Ploss and Max Bartels, *Das Weib in der Natur- und Völkerkunde*, ed. Paul Bartels, 2 vols. (Leipzig: Th. Grieben, 1908). With the exclusion of the added images, the text I quote here is the same in 1908 as in Ploss's 1885 version.

120. The German terminology for "women's language" changes in the later nineteenth century from *Weibersprache* to *Frauensprache* as the word *Frau* is substituted for *Weib* in everyday usage.

121. Ploss and Bartels, *Das Weib*, 1:199.

122. Ploss and Bartels, *Das Weib*, 1:202.

123. Ploss and Bartels, *Das Weib*, Vorrede zur ersten Auflage, ix.

124. Miyako Inoue, *Vicarious Language: Gender and Linguistic Modernity in Japan* (Berkeley: University of California Press, 2006), 14.

125. Friedrich Nietzsche, "On Truth and Lies in an Nonmoral Sense (1873)," in *The Nietzsche Reader*, ed. Keith Ansell Pearson and Duncan Large (Malden, MA: Blackwell, 2006), 114–23, here 116.

126. As Ute Gerhard summarizes, for some feminist groups, the *Frauenfrage* was a question of culture, while for others it was an issue of rights and freedoms. Ute Gerhard, *Frauenbewegung und Feminismus: Eine Geschichte seit 1789* (Munich: C. H. Beck, 2009), 73. See also Ann Taylor Allen, *Feminism and Motherhood in Germany, 1800–1914* (New Brunswick, NJ: Rutgers University Press, 1991).

127. Lou Andreas-Salomé, "Der Mensch als Weib: ein Bild im Umriss," in *Die Erotik: Vier Aufsätze*, ed. Ernst Pfeiffer (Frankfurt a.M.: Ullstein, 1992), 7–44, here 10. Muriel Cormican shows how *femininity* and *woman* are coterminous in this essay. See Cormican, *Women in the Works of Lou Andreas-Salomé: Negotiating Identity* (Rochester, NY: Camden House, 2009), 16 and 15.

128. Andreas-Salomé, "Der Mensch als Weib," 26. Andreas-Salomé also suggests a new understanding of woman's role in sexual reproduction, arguing—contrary to the traditional Aristotelian view—that the egg plays a more active role than the sperm.

129. See Ploss and Bartels, *Das Weib*, 1:58.

130. Ploss and Bartels, *Das Weib*, 1:58.

131. Ploss and Bartels, *Das Weib*, 1:58. This is a quotation from Runge.

132. Charles Darwin, *The Descent of Man, and Selection in Relation to Sex* (London: John Murray, 1871), 2:327.

133. Darwin, *Descent*, 2:326–27.

134. Nancy Leys Stepan, "Race and Gender: The Role of Analogy in Science," *Isis* 77, no. 2 (June 1986): 261–77, here 263.

135. Stepan, "Race and Gender," 263.

136. Carl Vogt, *Vorlesungen über den Menschen, seine Stellung in der Schöpfung und in der Geschichte der Erde* (Gießen: J. Ricker, 1863), 94. Quoted in Stepan, "Race and Gender," 263. Vogt furthermore claims that the differences between the sexes are more pronounced the more "civilized" a culture is—an idea that Darwin also quotes in *The Descent of Man*. See Vogt, *Vorlesungen*, 94–95, and Darwin, *Descent*, 2:330.

137. See Hermann Welcker, *Untersuchungen über Wachstum und Bau des menschlichen Schädels* (Leipzig: Wilhelm Engelmann, 1862).

138. Londa Schiebinger, *The Mind Has No Sex? Women in the Origins of Modern Science* (Cambridge, MA: Harvard University Press, 1991), 206.

139. See Schiebinger, *Nature's Body*, 158–72.

140. Georg von der Gabelentz, *Die Sprachwissenschaft: Ihre Aufgaben, Methoden und bisherigen Ergebnisse*, ed. Manfred Ringmacher and James McElvenny (Berlin: Language Science Press, 2016), 15. This edition brings together the 1891 and (posthumously published) 1901 editions of Gabelentz' *Sprachwissenschaft*, marking in the text what Gabelentz' 1901 editor, Albrecht Graf von Schulenburg, added to and subtracted from the second version. Unless otherwise indicated, all of my citations of Gabelentz' work were included in the original 1891 edition.

141. Gabelentz, *Sprachwissenschaft*, 15.

142. Gabelentz, *Sprachwissenschaft*, 15.

143. Gabelentz, *Sprachwissenschaft*, 18.

144. Gabelentz names Humboldt as an important founder of the discipline, but also criticizes him as being "unfortunately mostly less clear than deep, judges more than teaches, and presupposes knowledge that, for external reasons, few people can reach." Gabelentz, *Sprachwissenschaft*, 30.

145. Gabelentz, *Sprachwissenschaft*, 18.

146. Gabelentz, *Sprachwissenschaft*, 29.

147. Gabelentz, *Sprachwissenschaft*, 30.

148. Gabelentz, *Sprachwissenschaft*, ii.

149. Gabelentz, *Sprachwissenschaft*, ii.

150. Gabelentz, *Sprachwissenschaft*, 261.

151. Gabelentz, *Sprachwissenschaft*, 260.

152. Gabelentz, *Sprachwissenschaft*, 260, my emphasis.

153. See Schiebinger, *Nature's Body*, 99–106.

154. Gabelentz includes references to some scholarship in the 1891 version; even more detail was added by Albert Graf von der Schulenburg to the 1901 edition. See Gabelentz, *Sprachwissenschaft*, 261.

155. See Linda Ben-Zvi, "Samuel Beckett, Fritz Mauthner, and the Limits of Language," *PMLA* 95, no. 2 (March 1980): 183–200, here 187.

156. Fritz Mauthner, *Beiträge zu einer Kritik der Sprache* (Leipzig, 1923; repr., Hildesheim: Georg Olms, 1969), 1:2.

157. Mauthner, *Beiträge*, 1:2.

158. Katherine Arens, *Functionalism and Fin de Siècle: Fritz Mauthner's Critique of Language* (New York: Peter Lang, 1984), 63. See also Elizabeth Bredeck, *Metaphors of Knowledge: Language and Thought in Mauthner's Critique* (Detroit: Wayne State University Press, 1992).

159. On Mauthner's relationship to previous philosophical traditions, see Bredeck, *Metaphors*, 69.

160. Mauthner, *Beiträge*, 1:55.

161. This observation is made about numerous animals: for instance, the quail, the horse, the donkey, and the ox. About the horse, Buffon's text reads: "From birth the male has a stronger voice than the female; at puberty the voice of males and females becomes louder & deeper, as in humans and in most other animals." Le Comte de Buffon, *Œuvres completes: Tome premier, Histoire des Animaux quadrupèdes* (Paris: De l'Imprimerie Royal, 1775), 125.

162. Gaëtan Delaunay, *Études de biologie comparée basées sur l'évolution organique* (Paris: V. Adrien Delahaye, 1878), 1:99–100. Delaunay actually proposes the theory that it was the difference between men's and women's voices which led to masculine and feminine noun endings.

163. Darwin, *Descent*, 2:330.

164. Mauthner, *Beiträge*, 3:30.

165. Mauthner, *Beiträge*, 1:55.

166. Mauthner, *Beiträge*, 1:58.

167. Mauthner, *Beiträge*, 1:57.

168. Mauthner, *Beiträge*, 1:3.

169. Mauthner, *Beiträge*, 1:58, my emphasis.

170. Mauthner, *Beiträge*, 1:58–59.

171. See Mauthner, *Beiträge*, 1:60.

172. See Mauthner, *Beiträge*, 1:59.

173. Mauthner's main concern in this section is to compare human and animal sex drives, which he does in detail. See Mauthner, *Beiträge*, 1:45.

174. Bredeck, *Metaphors*, 100.

175. Mauthner, *Beiträge*, 1:61.

176. There is one notable exception to this: James G. Frazer's essay, "A Suggestion as to the Origin of Gender in Language," *Fortnightly Review* 73 (January 1900): 79–90. Frazer argues that men's/women's languages are potentially the source of grammatical gender.

177. Alexander F. Chamberlain, "Women's Languages," *American Anthropologist* 14, no. 3 (July–September 1912): 579–81, here 579.

178. See Richard Lasch, "Über Sondersprachen und ihre Entstehung," *Mitteilungen der Anthropologischen Gesellschaft in Wien* 37 (1907): 89–101. Already in 1884, the Swiss anthropologist Otto Stoll had compared the women's language of the Island Caribs with Arawak and concluded they are not related and that another explanation had to be uncovered. The French linguist Lucien Adam, however, had also compared the women's language with Arawak in 1879 and came up with the opposite result. See Lucien Adam, "Du parler des hommes et des femmes dans la langue caraïbe," *Revue de linguistique et de philologie comparée* xii (1879): 275–304. See also Otto Stoll, *Zur Ethnographie der Republik Guatemala* (Zürich: Orell Füssli, 1884).

179. "For thousands of years the work that especially fell to men was such as demanded an intense display of energy for a comparatively short period, mainly in war and in hunting. Here, however, there was not much occasion to talk, nay, in many circumstances talk might even be fraught with danger. [...] Woman, on the other hand, had a number of domestic occupations which did not claim such an enormous output of spasmodic energy. To her was at first left [...] the care of the children, cooking, brewing, baking, sewing, washing, etc.,—things which for the most part demanded no deep thought, which were performed in company and could well be accompanied with a lively chatter. Lingering effects of this state of things are seen still [...]" Otto Jespersen, "The Woman," in *Language: Its Nature, Development and Origin* (New York: Henry Holt, 1922), 237–54, here 254.

180. Flora Kraus, "Die Frauensprachen bei den primitiven Völkern," *Imago: Zeitschrift für Anwendung der Psychoanalyse auf die Geisteswissenschaften* 10 (1924): 296–313.

Chapter Three

1. Humboldt, "Ueber den Dualis [1827]," in *Werke in fünf Bänden*, ed. Andreas Flitner and Klaus Giel (Darmstadt: Wissenschaftliche Buchgesellschaft, 2002), 3:113–43.

2. Jacob Grimm, *Deutsche Grammatik*, vol. 3 (Göttingen: Dieterich, 1831), 403. All future references to this volume and edition will be given parenthetically as *DG*.

3. Anthony Corbeill, *Sexing the World: Grammatical Gender and Biological Sex in Ancient Rome* (Princeton, NJ: Princeton University Press, 2015).

4. Charles Francis Hockett, *A Course in Modern Linguistics* (New York: Macmillan, 1958), 231.

5. Hockett, *Course in Modern Linguistics*, 231.

6. For more on gender in English, and the problem of referring to it as a purely "natural" system, see Sally McConnell-Ginet, "Gender and Its Relation to Sex: The Myth of 'Natural' Gender," in *The Expression of Gender*, ed. Greville G. Corbett (Berlin: de Gruyter Mouton, 2014), 3–38.

7. This summary of historical theories of grammatical gender is indebted to the overview in Gerlach Royen, *Die nominalen Klassifikations-Systeme in den Sprachen der Erde* (Mödling: Anthropos, 1929). Although most grammarians during this period were chiefly concerned with

cataloging and understanding their native languages, a number of grammatical studies circulated across national borders. Girard and Harris, for example, were read in Germany, Harris's text even before it was translated into German (Royen, *Klassifikations-Systeme*, 30).

8. Antoine Arnauld and Claude Lancelot, *Grammaire générale et raisonnée de Port-Royal* (Paris: Perlet, 1803), 282. Antoine Arnauld and Claude Lancelot, *General and Rational Grammar: The Port-Royal Grammar*, ed. and trans. by Jacques Rieux and Bernard E. Rollin (The Hague: Mouton, 1975), 77.

9. J. G. Schottelius, *Teutsche Sprachkunst* (Braunschweig: Gruber, 1641), 249.

10. Gabriel Girard; *Les vrais principes de la langue Françoise: ou, La parole réduite en méthode, conformément aux loix de l'usage, en seize discours* (Paris: Le Breton, 1747), 1:160.

11. Girard, *Les vrais principes*, 1:225.

12. Girard, *Les vrais principes*, 1:225–26.

13. Nicolas Beauzée and Jacques-Philippe-Augustin Douchet, "Genre (Grammaire)," in *Encyclopédie, ou dictionnaire raisonné des sciences, des arts et des métiers, etc.*, ed. Denis Diderot and Jean le Rond d'Alembert. University of Chicago ARTFL Encyclopédie Project, 2016, ed. Robert Morrissey and Glenn Roe, 590; accessed June 1, 2019, http://encyclopedie.uchicago.edu.

14. Beauzée and Douchet, "Genre."

15. Johann Christoph Gottsched, *Grundlegung einer deutschen Sprachkunst* (Leipzig: Bernhard Christoph Breitkopf, 1748), 160.

16. Gottsched, *Grundlegung*, 167.

17. Antoine Court de Gébelin, *Monde primitif, analysé et comparé avec le monde moderne, considéré dans l'histoire naturelle de la parole* (Paris: Boudet, 1774), 2:72.

18. Court de Gébelin, *Monde primitif*, 2:74.

19. Johann Christoph Adelung, *Deutsche Sprachlehre* (Berlin: Christian Friedrich Voß und Sohn, 1781), 116.

20. August Ferdinand Bernhardi, *Sprachlehre* (Berlin: Heinrich Frölich, 1801–3), 1:143.

21. Ernst Cassirer, "The Form of the Concept in Mythical Thinking (1922)," in *The Warburg Years (1919–1933): Essays on Language, Art, Myth, and Technology*, trans. S. G. Lofts with A. Calcagno (New Haven, CT: Yale University Press, 2013), 1–71, here 10.

22. Corbeill, *Sexing the World*, 13.

23. Dennis Baron, *Grammar and Gender* (New Haven, CT: Yale University Press, 1986), 91.

24. Herder, *Abhandlung*, 738; *Treatise*, 101.

25. Adelung published three texts that deal with grammatical gender; each text built upon the previous one. The first discussion of grammatical gender, from 1781, was published in Adelung's *Deutsche Sprachlehre*. The second, "Von dem Geschlechte der Hauptwörter," was published in Adelung's periodical, *Umständliches Lehrgebäude der Deutschen Sprache, zur Erläuterung der Deutschen Sprachlehre für Schulen*, in 1782. In 1783, Adelung then published "Von dem Geschlechte der Substantive" in his *Magazin für die deutsche Sprache*.

26. Adelung, "Von dem Geschlechte der Substantive," *Magazin für die deutsche Sprache* 1, no. 4 (1783): 3–20, here 11.

27. Adelung, "Geschlechte der Substantive," 9.

28. Adelung, "Geschlechte der Substantive," 19.

29. Adelung, "Geschlechte der Substantive," 12.

30. Adelung, "Geschlechte der Substantive," 12–13.

31. Already in the 1782 revised edition of his first essay on gender (from 1781), Adelung begins to conceive of the neuter as the absence of sex. He criticizes scholars who call neuter *das*

Ungewisse, since the neuter should stand for lack of sex, not unclear sex. In this essay, however, Adelung has not yet developed the theory that the neuter dates from a later period than the masculine and feminine. Johann Christoph Adelung, "Von dem Geschlechte der Hauptwörter," in *Umständliches Lehrgebäude der Deutschen Sprache, zur Erläuterung der Deutschen Sprachlehre für Schulen* (Leipzig: Breitkopf, 1782), 1:343–69, here 345.

32. Adelung, "Von dem Geschlechte der Hauptwörter," 344.

33. August Ferdinand Bernhardi, *Anfangsgründe der Sprachwissenschaft* (Berlin: Heinrich Frölich, 1805), 131.

34. Karl Philipp Moritz, *Deutsche Sprachlehre für die Damen: In Briefen* (Berlin: Arnold Wever, 1782), 231.

35. Moritz, *Deutsche Sprachlehre für die Damen*, 230–31.

36. Lann Hornscheidt, *feministische w_orte: ein lern-, denk- und handlungsbuch zu sprache und diskriminierung, gender studies und feministischer linguistic* (Frankfurt a.M.: Brandes & Apsel, 2012), 76.

37. See Johann Werner Meiner, *Versuch einer an der menschlichen Sprache abgebildeten Vernunftlehre: oder Philosophische und allgemeine Sprachlehre* (Leipzig: Breitkopf, 1781) and Christian Heinrich Wolke, *Anleit zur deutschen Gesamtsprache oder zur Erkennung und Berichtung einiger (zu wenigst 20) tausend Sprachfehler in der hochdeutschen Mundart* (Dresden, 1812). As discussed in Royen, *Klassifikations-Systeme*, 32 and 37.

38. Wilhelm von Humboldt, *Lettre à m. Abel-Rémusat, sur la nature des formes grammaticales en général, et sur le génie de la langue chinoise en particulier* (Paris: Dondey-Dupré, 1827), 12–13, my emphasis.

39. Humboldt, "Dualis," 134–35.

40. Humboldt, "Dualis," 136.

41. Humboldt, "Dualis," 137.

42. Humboldt, "Dualis," 138.

43. Humboldt, "Dualis," 140.

44. Humboldt, "Dualis," 141.

45. Humboldt, "Dualis," 141.

46. August Pott, "Metaphern, vom leben und von körperlichen lebensverrichtungen hergenommen," *Zeitschrift für vergleichende Sprachforschung auf dem Gebiete des Deutschen, Griechischen und Lateinischen* 2 (1853): 101–27, here 103.

47. Humboldt, "Dualis," 138.

48. Römer, *Sprachwissenschaft und Rassenideologie*, 124.

49. Benes, *Babel's Shadow*, 211.

50. Bleek, *Ursprung*, xvi; *Origin of Language*, xxii.

51. Bleek, *Ursprung*, xix; *Origin*, xxvi. That speakers of languages with grammatical gender have higher "mental tendencies" is an idea that Bleek developed in earlier works, including his 1851 dissertation at the University of Bonn and his 1862 *A Comparative Grammar of South African Languages* (London: Trübner, 1862). See Andrew Bank, "Evolution and Racial Theory: The Hidden Side of Wilhelm Bleek," *South African History Journal* 43 (November 2000), 163–78, here 172.

52. Bleek, *Ursprung*, xvi; *Origin*, xxiii, translation modified.

53. Bleek, *Ursprung*, xv; *Origin*, xxiii.

54. Robert Bernasconi, "Who Invented the Concept of Race? Kant's Role in the Enlightenment Construction of Race," in *Race*, ed. Robert Bernasconi (Malden, MA: Blackwell, 2001), 11–36, here 11. I adhere here to Bernasconi's explanation of a "scientific concept": a concept "with

sufficient definition for subsequent users to believe that they were addressing something whose scientific status could at least be debated" (11).

55. Other texts include the 1775 "Of the Different Human Races" and the 1788 "On the Use of Teleological Principles in Philosophy."

56. Immanuel Kant, "Of the Different Human Races (1777)," in *Kant and the Concept of Race: Late Eighteenth-Century Writings*, ed. Jon M. Mikkelsen (Albany: State University of New York, 2013), 55–71, here 62.

57. Kant, "Of the Different Human Races (1777)," 67.

58. See Bank, "Evolution and Racial Theory," as well as Rachael Gilmour, "From Languages to Language: The Comparative Philologist in South Africa," in *Grammars of Colonialism: Representing Languages in Colonial South Africa* (New York: Palgrave Macmillan, 2007), 169–91.

59. Bleek, *Ursprung*, v; *Origin*, xi.

60. Bleek, *Ursprung*, iv–v; *Origin*, xi.

61. Bleek, *Ursprung*, iv–v; *Origin*, x.

62. Bleek, *Ursprung*, xiv; *Origin*, xx.

63. Stefan Arvidsson, *Aryan Idols: Indo-European Mythology as Ideology and Science*, trans. Sonia Wichmann (Chicago: University of Chicago Press, 2006), 82. See volume 2 of Max Müller's *Chips from a German Workshop* (London: Longmans, Green, and Co., 1867), 55–56.

64. On Bleek's encounter with Lepsius, see *Claim to the Country: The Archive of Lucy Lloyd and Wilhelm Bleek*, ed. Pippa Skotnes (Johannesburg: Jacana, 2007), 184.

65. Karl Richard Lepsius, *Standard Alphabet for Reducing Unwritten Languages and Foreign Graphic Systems to a Uniform Orthography in European Letters*, 2nd ed. (London: Williams & Norgate, 1863), 89. The first edition, which was published in German and in English in the 1850s, does not include this particular section. Referenced in Morpurgo Davies, *Nineteenth-Century Linguistics*, 158.

66. Lepsius, *Standard Alphabet*, 90.

67. Karl Richard Lepsius, *Nubische Grammatik* (Berlin: Wilhelm Hertz, 1880), xxii. This reference comes from Baron, *Grammar and Gender*, 92, translation modified.

68. This is also discussed in Benes, *Babel's Shadow*, 138.

69. Grimm, *Über den Ursprung der Sprache*, 43.

70. Willem Gerard Brill, *Hollandsche Spraakleer* (Leiden: Luchtmans, 1846), 294. Quoted in German translation in Royen, *Klassifikations-Systeme*, 56.

71. Philipp Franz von Walther, *Physiologie des Menschen mit durchgängiger Rücksicht auf die comparative Physiologie der Thiere*, vol. 2 (Landshut: Krüll, 1808), 373-374. As quoted in Honegger, *Ordnung*, 188.

72. Walther, *Physiologie*, 375.

73. Jacob Grimm and Wilhelm Grimm, *Deutsches Wörterbuch* (Leipzig: S. Hirzel, 1854), 1:xiii. For more on this, see Michael Townson, *Mother-Tongue and Fatherland: Language and Politics in German* (Manchester: Manchester University Press, 1992), 94.

74. Grimm and Grimm, *Wörterbuch*, 1:xiii.

75. Jakob Norberg, *The Brothers Grimm and the Making of German Nationalism* (Cambridge: Cambridge University Press, 2022), 144.

76. Jacob Grimm, *Kleinere Schriften*, vol. 1, ed. Karl Müllenhoff (Berlin: Ferd. Dümmler, 1864), 224. Quoted and translated in Norberg, *Brothers Grimm*, 150.

77. Jacob Grimm and Wilhelm Grimm, "Preface to Volume II," in *The Original Folk and Fairy Tales of the Brothers Grimm: The Complete First Edition*, trans. and ed. Jack Zipes (Princeton, NJ: Princeton University Press, 2015), 269–73, here 270.

78. Benes, *Babel's Shadow*, 126.

79. Jacob Grimm, "Einige Hauptsätze, die ich aus der Geschichte der deutschen Sprache gelernt habe," in *Kleinere Schriften* (Hildesheim: Olms, 1965–66), 8:45–55, here 45. This text was originally part of the preface to volume 1 of *Deutsche Grammatik* and is also discussed in John Edward Toews, *Becoming Historical: Cultural Reformation and Public Memory in Early Nineteenth-Century Berlin* (Cambridge: Cambridge University Press, 2004), 345.

80. Jacob Grimm, *Deutsche Grammatik*, vol. 3, ed. Gustav Roethe and Edward Schröder (Gütersloh: C. Bertelsmann, 1890), 319. This question was added by a later editor but is in line with Grimm's original claim (from the 1831 edition) about the division of male and female proper names (*DG* 323).

81. Toews, *Becoming Historical*, 363.

82. Orrin W. Robinson, *Grimm Language: Grammar, Gender and Genuineness in the Fairy Tales* (Philadelphia: John Benjamins, 2010), 85 and 172.

83. Jack Zipes, *Fairy Tales and the Art of Subversion: The Classical Genre for Children and the Process of Civilization* (New York: Routledge, 1991), 58–59.

84. Leopold von Ranke, the so-called "father" of the discipline of history, is even more explicit in his sexualization of the archive, as Bonnie Smith shows. In an 1827 letter, Ranke writes of an Italian archive, "Yesterday I had a sweet, magnificent fling with the object of my love, a beautiful Italian, and I hope that we produce a beautiful Roman-German prodigy." Smith, *The Gender of History*, 119.

85. Kilarski, *Nominal Classification*, 128.

86. See Karl Brugmann, *On the Nature and Origin of the Noun Genders in the Indo-European Languages*, trans. Edmund Y. Robbins (New York: Charles Scribner's Sons, 1897), 32.

87. Kilarski, *Nominal Classification*, 129–31.

88. Hermann Paul, *Principien der Sprachgeschichte* (Halle: Max Niemeyer, 1886), 220; *Principles of the History of Language*, trans. H. A. Strong (London: Swan Sonnenschein, Lowrey, & Co., 1888), 289.

89. Paul, *Principien*, 220; *Principles*, 289.

90. Gustav Roethe, "Zum neuen Abdruck," in *Deutsche Grammatik*, vol. 3, by Jacob Grimm (Gütersloh: Bertelsmann, 1890), ix–xxxi, here xxiv.

91. Karl Brugmann, "Zur Frage der Entstehung des grammatischen Geschlechts," *Beiträge zur Geschichte der deutschen Sprache und Literatur* 15 (1891): 523–31, here 529.

92. Karl Brugmann, "Das Nominalgeschlecht in den indogermanischen Sprachen," *Internationale Zeitschrift für allgemeine Sprachwissenschaft* 4 (1889): 100–109, here 104.

93. Brugmann, *Nature and Origin*, 27; discussed in Kilarski, *Nominal Classification*, 135.

94. Brugmann, *Nature and Origin*, 31.

Chapter Four

1. Virginia Woolf, *A Room of One's Own*, in *A Room of One's Own* and *Three Guineas*, ed. Morag Shiach (Oxford: Oxford University Press, 1992), 1–149, here 61.

2. Woolf, *Room of One's Own*, 62.

3. Annemete von Vogel and James McElvenny, "The Gabelentz Family in Their Own Words," in *Georg von der Gabelentz and the Science of Language*, ed. James McElvenny (Amsterdam: Amsterdam University Press, 2019), 13–26, here 16.

4. Vogel and McElvenny, "Gabelentz Family," 15.

5. This is according to the biography provided in the exhibition catalogue, *Nadelkunst der Clementine von Münchhausen, 1849–1913* (Wunstorf: A. Jacques, 2000), 2.

6. *Nadelkunst der Clementine von Münchhausen*, 2–3.

7. Börries von Münchhausen, "Clementine von Münchhausen geb. v. d. Gabelentz: Anlagen und Talente, Entwicklung u. Studien" (unpublished manuscript, 1907), typescript.

8. Clementine von Münchhausen, "H. Georg v. d. Gabelentz. Biographie und Charakteristik [1913]," ed. Annemete von Vogel, in *Georg von der Gabelentz: Ein biographisches Lesebuch*, ed. Kennosuke Ezawa and Annemete von Vogel (Tübingen: Gunter Narr, 2013), 85–171, here 114.

9. Münchhausen, "Biographie und Charakteristik," 125.

10. Münchhausen, "Biographie und Charakteristik," 144.

11. Münchhausen, "Biographie und Charakteristik," 143.

12. Münchhausen, "Biographie und Charakteristik," 147.

13. Münchhausen, "Biographie und Charakteristik," 167.

14. Münchhausen, "Biographie und Charakteristik," 141.

15. Münchhausen, "Biographie und Charakteristik," 123.

16. Münchhausen, "Biographie und Charakteristik," 141.

17. Sara Ahmed, *Queer Phenomenology: Orientations, Objects, Others* (Durham, NC: Duke University Press, 2006), 30.

18. Salomon Lefmann, *August Schleicher: Skizze* (Leipzig: B. G. Teubner, 1870), 91.

19. Lefmann, *Schleicher*, 34.

20. Lefmann, *Schleicher*, 34.

21. Saidiya Hartman, "Venus in Two Acts," *Small Axe* 12, no. 2 (June 2008): 1–14, here 11.

22. Albisetti, *Schooling German Girls and Women*, 285.

23. For a list of members and contributors, see E. F. K. Koerner, *The Importance of Techmer's Internationale Zeitschrift für Allgemeine Sprachwissenschaft in the Development of General Linguistics: An Essay* (Amsterdam: John Benjamins, 1973).

24. Wendy Ayres-Bennett and Helena Sanson, "Women in the History of Linguistics: Distant and Neglected Voices," in *Women in the History of Linguistics*, 1–30. The editors make this case about Romania, for instance (11).

25. Sabrina Ebbersmeyer, "From a 'Memorable Place' to 'Drops in the Ocean': On the Marginalization of Women Philosophers in German Historiography of Philosophy," *British Journal for the History of Philosophy* 28, no. 3 (May 2020): 442–62.

26. Nicola McLelland, "Women in the German Linguistic Tradition," in *Women in the History of Linguistics*, ed. Ayres-Bennett and Sanson, 193–217, here 207.

27. Alan Corkhill, "Female Language Theory in the Age of Goethe: Three Case Studies," *Modern Language Review* 94, no. 4 (October 1999): 1041–53. Corkhill essentializes women's relation to language, arguing that "there exists an undeniable gender-driven *need* to reflect on language as an integral part of the process of female self-definition" (1053).

28. Sarah Tyson, *Where are the Women? Why Expanding the Archive Makes Philosophy Better* (New York: Columbia University Press, 2018), xxxiii.

29. James C. Albisetti, "German Influence on the Higher Education of American Women, 1865–1914," in *German Influences on Education in the United States to 1917*, ed. Henry Geitz, Jürgen Heideking, and Jurgen Herbst (Cambridge: Cambridge University Press, 1995), 227–44; McLelland, "Women in the German Linguistic Tradition," 203–6.

30. Before beginning at Wellesley, Wenckebach trained at the "Sauveur School" at Amherst College. The director, Lambert Sauveur, taught a "conversation-based method dependent on the

teacher's ability to teach the meaning of new words by object lessons, pictures, mime, context, and so on," according to A. P. R. Howatt and Richard Smith, "The History of Teaching English as a Foreign Language, from a British and European Perspective," *Language and History* 57, no. 1 (May 2014): 75–95, here 83. Wenckebach also mentions Pestalozzi as an influence, for instance in the preface to *Deutscher Anschauungs-Unterricht für Amerikaner*.

31. See Kittler, *Discourse Networks*, 25–69.

32. Margarethe Müller, *Carla Wenckebach, Pioneer* (Boston: Ginn, 1908). Müller bases her biography on the multiple interviews she conducted with Wenckebach's family in Germany, as well as on an apparent "autobiographical manuscript novel" written by Wenckebach herself (ix).

33. This is also "the woman in whose mental make-up sex does not appear to be of prime and decisive importance" (Müller, *Carla Wenckebach, Pioneer*, vi).

34. Müller, *Carla Wenckebach, Pioneer*, 23.

35. Müller, *Carla Wenckebach, Pioneer*, 25.

36. Müller, *Carla Wenckebach, Pioneer*, 5.

37. Müller, *Carla Wenckebach, Pioneer*, 6.

38. Müller, *Carla Wenckebach, Pioneer*, 219.

39. In 1884, this book was first self-published and then republished by a commercial publisher. Here, I am referring to the fourth edition: Carla Wenckebach and Josepha Schrakamp, *Deutsche Grammatik für Amerikaner: Nach einer praktischen Methode* (New York: F. W. Christern, 1884), 91.

40. Wenckebach and Schrakamp, *Deutsche Grammatik für Amerikaner*, 122.

41. For example: "Der Vater verreist mit dem Sohn" (the father travels with the son; 157); "Welcher Schüler hat seinen Vater verloren?" (Which [male] student lost his father?). This relation to the father is not repeated in the following section on "Schülerin" (female student; 117).

42. Carla Wenckebach and Helene Wenckebach, *Deutscher Anschauungs-Unterricht für Amerikaner: Ein Hilfsbuch* (Boston: Carl Schoenhof, 1886).

43. Wenckebach and Wenckebach, *Deutscher Anschauungs-Unterricht*, iii.

44. Carla Wenckebach and Helene Wenckebach, *Deutsches Lesebuch* (Boston: Carl Schoenhof, 1887).

45. As Dennis Baron shows, there is a long tradition of English speakers using alternatives to *he* as a generic pronoun. There are records of the singular *they* as far back as the fourteenth century. Baron, *What's Your Pronoun*, 149.

46. Wenckebach and Wenckebach, *Deutsches Lesebuch*, xii.

47. Wenckebach and Wenckebach, *Deutsches Lesebuch*, ix.

48. Carolina Michaëlis de Vasconcellos, *Studien zur romanischen Wortschöpfung* (Leipzig: F. A. Brockhaus, 1876), vii.

49. Michaëlis de Vasconcellos, *Studien*, viii and 11.

50. Michaëlis de Vasconcellos, *Studien*, 10.

51. Michaëlis de Vasconcellos, *Studien*, 1.

52. Michaëlis de Vasconcellos, *Studien*, 1.

53. J. H. Oswald, *Das grammatische Geschlecht und seine sprachliche Bedeutung* (Paderborn: Junfermann, 1866), 61. For more on the history of the term *Tochtersprache* (daughter language), see Georgia Veldre, "Zur Diskussion über den Begriff 'Tochtersprache' im 19. Jahrhundert," *Historiographia Linguistica* 19, no. 1 (1992): 65–96.

54. Albisetti, *Schooling German Girls and Women*, 18.

55. F. H. C. Schwarz, *Grundsätze der Töchtererziehung für die Gebildeten* (Jena: Cröker, 1836), 152. Quoted in Albisetti, *Schooling German Girls and Women*, 119.

56. Schwarz, *Grundsätze*, 152.

57. Michaëlis de Vasconcellos, *Studien*, 8.

58. Michaëlis de Vasconcellos, *Studien*, vii and 49.

59. Michaëlis de Vasconcellos, *Studien*, 9.

60. Michaëlis de Vasconcellos, *Studien*, 10–11.

61. Heinz Kröll, "Michaëlis de Vasconcellos, Carolina," in *Neue Deutsche Biographie* 17 (1994): 437–38, online ed., https://www.deutsche-biographie.de/pnd119282585.html#ndbcontent.

62. See Albisetti, *Schooling German Girls and Women*, 16.

63. Sónia Coelho, Susana Fontes, and Rolf Kemmler, "The Female Contribution to Language Studies in Portugal," in *Women in the History of Linguistics*, ed. Ayres-Bennett and Sanson, 145–66, here 158–59.

64. Michaëlis de Vasconcellos, *Studien*, vii.

65. Michaela Raggam-Blesch, "A Pioneer in Academia: Elise Richter," in *Jewish Intellectual Women in Central Europe, 1860–2000: Twelve Biographical Essays*, ed. Judith Szapor, Andrea Petö, and Marina Calloni (Lewiston, NY: Edwin Mellen, 2012), 93–128.

66. Elise Richter, "Erziehung und Entwicklung," in *Kleinere Schriften zur allgemeinen und romanischen Sprachwissenschaft*, ed. Yakov Malkiel (Innsbruck: H. Kowatsch, 1977), 531–54, here 538. This autobiographical essay was first published in *Führende Frauen Europas in sechzehn Selbstschilderungen*, ed. Elga Kern (Munich: E. Reinhardt, 1928), 70–93.

67. Richter, "Erziehung," 543.

68. Richter, "Erziehung," 546.

69. See Raggam-Blesch, "Pioneer," 110.

70. For a comprehensive overview of Richter's scholarly output, see the bibliography by B. M. Woodbridge Jr. in Richter, *Kleinere Schriften*, 583–99.

71. In 1866, the Linguistic Society of Paris famously banned discussions of the origin of language. As James Stam writes, this did not stop people from producing texts on the topic, but it did indicate a general turn away from focusing on the origin of language in linguistic discourse. See Stam, "The Annihilation of the Question," in *Inquiries into the Origin of Language*, 255–62.

72. E. F. K. Koerner, "Historiography of Phonetics: The State of the Art," *Journal of the International Phonetic Association* 23, no. 1 (June 1993): 1–12. Koerner makes this case via Panconcelli-Calzia's 1941 *Geschichtszahlen der Phonetik: 3000 Jahre Phonetik*.

73. For an overview of the instruments of nineteenth-century experimental phonetics, see Sylvain Auroux, E. F. K. Koerner, Hans-Josef Niederehe, and Kees Versteegh, *History of the Language Sciences / Geschichte der Sprachwissenschaften / Histoire des Sciences du langage* (Berlin: de Gruyter Mouton, 2001), 2:1470–74. For a general overview, see also K. Kohler, "Three Trends in Phonetics: The Development of Phonetics as a Discipline in Germany since the Nineteenth Century," in *Towards a History of Phonetics*, ed. R. E. Asher and Eugénie J. A. Henderson (Edinburgh: Edinburgh University Press, 1981), 161–78.

74. Tobias Wilke, *Sound Writing: Experimental Modernism and the Poetics of Articulation* (Chicago: University of Chicago Press, 2022), 4–5.

75. In Techmer's case, there is one exception: an image of a woman operating a talking machine created by Joseph Faber. This image of Faber's euphonia circulated widely and is not Techmer's own creation. As for his own sources: Techmer writes that he mainly made use of his publisher's wood carvings, while also copying images from anatomical works by Jakob Henle and books on sound by Hermann Helmholtz. See the "Vorbemerkungen" to volume 2 of *Phonetik*

(no pagination). In addition to the two authors he mentions in the preface, Techmer also borrowed from a range of other physiological and anatomical sources.

76. Friedrich Techmer, *Phonetik: Zur vergleichenden Physiologie der Stimme und Sprache*, 2 vols. (Leipzig: Wilhelm Engelmann, 1880). Techmer's source for this series of diagrams is the physiologist Johann Nepomuk Czermak. See Czermak, *Gesammelte Schriften* (Leipzig: Wilhelm Engelmann, 1879), 2:81.

77. Elise Richter, *Wie wir sprechen: Sechs volkstümliche Vorträge* (Leipzig: B. G. Teubner, 1912).

78. Christopher Oldstone-Moore, *Of Beards and Men: The Revealing History of Facial Hair* (Chicago: University of Chicago Press, 2015), 220.

79. Her images of the larynx (which she borrows from other texts) bear a striking resemblance to the female genitals, but, as in other phonetics texts of her era, Richter does not assert any such connection. Zelda Boyd and Thomas Laqueur have pointed to an image of the larynx in Max Müller's 1861 *Lectures on the Science of Language* that looks especially like a vagina—although, notably, this is not a connection that Müller's text itself makes. In ancient and early modern science, there is a more explicit tradition of associating the larynx with the female genitals, including Galen's argument that (in Laqueur's paraphrase) the uvula "gives the same sort of protection to the throat that the clitoris gives to the uterus." See Thomas Walter Laqueur, *Making Sex: Body and Gender from the Greeks to Freud* (Cambridge, MA: Harvard University Press, 1992), 37; and Zelda Boyd, "'The Grammarian's Funeral' and the Erotics of Grammar," *Browning Institute Studies* 16 (1988): 1–14.

80. Richter, *Wie wir sprechen*, 1.

81. Quoted in Christiane Hoffrath, *Bücherspuren: Das Schicksal von Elise und Helene Richter und ihrer Bibliothek im "Dritten Reich"* (Cologne: Böhlau, 2009), 48.

82. See Utz Maas, "Die erste Generation der deutschsprachigen Sprachwissenschaftlerinnen," *STUF—Language Typology and Universals* 44, no. 1 (December 1991): 61–69; McLelland, "Women in the German Linguistic Tradition," 216–17.

83. See Jacques R. Pauwels, *Women, Nazis, and Universities: Female University Students in the Third Reich, 1933–1945* (Westport, CT: Greenwood Press, 1984). Nazi ideology excluded women from higher education out of a belief that "women were intellectually unfit for higher studies and in any case not genuinely interested in such studies" (14) and out of fear that letting women study at university would negatively influence the German birth rate (15). Pauwels's work also shows, however, that the decline of the female student population during the Nazi period cannot be solely attributed to Nazi opposition (in fact, many of the party's antifeminist policies were ineffective in practice) but is rather the result of complex economic and demographic changes, as well as a general ideology of anti-intellectualism (*Women, Nazis, and Universities*, 45).

84. Michèle Le Dœuff, "Epilogue (2006)," in *Hipparchia's Choice: An Essay Concerning Women, Philosophy, etc.*, trans. Trista Selous (New York: Columbia University Press, 2007), 317–21, here 319. Le Dœuff is referring to *An Essay Concerning Human Understanding*, where Locke offers the story of Sempronia, who dug her sons out of a parsley bed, but whose sons were nonetheless brothers, as an example of how "we have ordinarily a clear (or clearer) Notion of the Relation, as of its Foundation." As Locke writes, "If I know what it is for one Man to be born of a Woman, viz. Sempronia, I know what it is for another Man to be born of the same Woman, Sempronia; and so have as clear a Notion of Brothers, as of Births, and, perhaps, clearer." See Locke, *Human Understanding*, 361.

Chapter Five

1. Rudolf Helmstetter, for instance, argues vehemently against using the term *Sprachkrise* in reference to Hofmannsthal. Helmstetter, "Entwendet: Hofmannsthals *Chandos-Brief*, die Rezeptionsgeschichte und die Sprachkrise," *Deutsche Vierteljahrsschrift für Literaturwissenschaft und Geistesgeschichte* 77, no. 3 (2003): 446–80.

2. Formigari, "Idealism and Idealistic Trends," 234.

3. Ferdinand de Saussure, *Course in General Linguistics*, trans. Wade Baskin, ed. Perry Meisel and Haun Saussy (New York: Columbia University Press, 2011).

4. Pourciau, *The Writing of Spirit*, 76–77.

5. Ludwig Wittgenstein, *Tractatus Logico-Philosophicus*, ed. Luciano Bazzocchi (London: Anthem Press, 2022), 155. Cora Diamond argues that for Wittgenstein, these limits are not necessarily negative: they can lead to satisfaction rather than resignation if we recognize that "there was *nothing* that we were asking or wanting to say." Cora Diamond, "The *Tractatus* and the Limits of Sense," in *The Oxford Handbook of Wittgenstein*, ed. Oskari Kuusela and Marie McGinn (Oxford: Oxford University Press, 2011): 240–75, here 240. In the literary texts examined in this chapter, however, the resignation incited by such limits of language is unambiguous.

6. Wittgenstein, *Tractatus*, 56.

7. Sigrid Weigel, *Body- and Image-Space: Re-Reading Walter Benjamin*, trans. Georgina Paul with Rachel McNicholl and Jeremy Gaines (London: Routledge, 1996), 86.

8. Carsten Zelle, "Konstellationen der Moderne. Verstummen—Medienwechsel—literarische Phänomenologie," *Musil-Forum* 27 (2001/2002): 88–102. Alys X. George, *The Naked Truth: Viennese Modernism and the Body* (Chicago: University of Chicago Press, 2020), 171. A similar argument is also made by Michael Hamburger in "Art as Second Nature: The Figures of the Actor and the Dancer in the Works of Hugo von Hofmannsthal," in *Romantic Mythologies*, ed. Ian Fletcher (London: Routledge, 1967), 225–41. George in particular shows that Hofmannsthal's so-called "crisis of language" did not begin with the Chandos letter but was in fact an interest from the beginning of his career.

9. Hugo von Hofmannsthal, "Ein Brief," *Erzählungen. Erfunden Gespräche und Briefe. Reisen*, ed. Bernd Schoeller, *Gesammelte Werke in zehn Einzelbänden* (Frankfurt a.M.: Fischer Taschenbuch, 1979), 461–72, here 472. Hugo von Hofmannsthal, "The Letter of Lord Chandos," in *Selected Prose*, trans. Mary Hottinger, Tania Stern, and James Stern (New York: Pantheon Books, 1952), 129–41, here 141.

10. George, *The Naked Truth*, 177.

11. The title of Lucian's text has also been variously translated as "On Dance" and "On Dancing," although Hofmannsthal writes in his essay that Lucian's text has the same title as his own. For an in-depth analysis of Lucian's dialogue, see Karin Schlapbach, "Lucian's *On Dancing* and the Models for a Discourse on Pantomime," in *New Directions in Ancient Pantomime*, ed. Edith Hall and Rose Wyles (Oxford: Oxford University Press, 2008), 314–37.

12. Hugo von Hofmannsthal, "Über die Pantomime," *Reden und Aufsätze I*, ed. Bernd Schoeller, *Gesammelte Werke in zehn Einzelbänden* (Frankfurt a.M.: Fischer Taschenbuch, 1979), 502–5, here 502.

13. Lucian writes, "As in literature, so too in dancing what is generally called 'bad taste' comes in when they exceed the due limit of mimicry and put forth greater effort than they should; if something dainty, they make it extravagantly effeminate, and they carry masculinity

to the point of savagery and bestiality." Lucian, *Lucian, Volume 5*, trans. A. M. Harmon, Loeb Classical Library 302 (Cambridge, MA: Harvard University Press, 1936), 285.

14. Hugo von Hofmannsthal, "Die unvergleichliche Tänzerin," *Reden und Aufsätze I*, 296–501, here 497.

15. For an in-depth account of Hofmannsthal's collaborations with female dancers, see "Hugo von Hofmannsthal and Grete Wiesenthal" in Mary Fleischer's *Embodied Texts: Symbolist Playwright-Dancer Collaborations* (Amsterdam: Rodopi, 2007), 93–148.

16. Hofmannsthal, "Pantomime," 502.

17. Hofmannsthal, *Reden und Aufsätze I*, 499.

18. Hofmannsthal, *Reden und Aufsätze I*, 497.

19. Hofmannsthal, *Reden und Aufsätze I*, 496.

20. Toni Morrison, *Playing in the Dark: Whiteness and the Literary Imagination* (Cambridge, MA: Harvard University Press, 1992), 68.

21. Morrison, *Playing in the Dark*, 36.

22. For a contemporary English translation of Lucian's text, see Lucian, *Lucian, Works*, vol. 7, trans. M. D. MacLeod, Loeb Classical Library 437 (Cambridge, MA: Harvard University Press, 1961).

23. Hugo von Hofmannsthal, "Furcht," in *Erzählungen. Erfundene Gespräche und Briefe. Reisen*, ed. Bernd Schoeller, *Gesammelte Werke in zehn Einzelbänden* (Frankfurt a.M.: Fischer Taschenbuch, 1979), 572–79, here 574; "Fear: A Dialogue," in *Selected Prose*, 155–64, here 157–58.

24. Hofmannsthal, "Furcht," 575; "Fear," 158.

25. Hofmannsthal, "Furcht," 577; "Fear," 161.

26. Gabriele Brandstetter, "Der Traum vom anderen Tanz: Hofmannsthals Ästhetik des Schöpferischen im Dialog 'Furcht,'" in *Hugo von Hofmannsthal: Neue Wege der Forschung*, ed. Elsbeth Dangel-Pelloquin (Darmstadt: Wissenschaftliche Buchgesellschaft, 2007), 41–61.

27. Hofmannsthal, "Furcht," 575; "Fear," 159.

28. Hofmannstahl, "Furcht," 575; "Fear," 159.

29. Hofmannsthal, "Furcht," 578; "Fear," 163.

30. See Brandstetter, "Der Traum vom anderen Tanz," 55. This is in reference to the essay by Rudolf Kassner, "Der indische Idealismus" (1903), which Hofmannsthal read and appreciated.

31. Hofmannsthal, "Furcht," 579; "Fear," 163.

32. For instance, Jill Scott, "From Pathology to Performance: Hugo von Hofmannsthal's *Elektra* and Sigmund Freud's 'Fräulein Anna O.,'" in *Electra after Freud: Myth and Culture* (Ithaca, NY: Cornell University Press, 2005), 57–80; Lorna Martens, "The Theme of the Repressed Memory in Hofmannsthal's *Elektra*," *German Quarterly* 60, no. 1 (Winter 1987): 38–51; Ritchie Robertson, "'Ich habe ihm das Beil nicht geben können': The Heroine's Failure in Hofmannsthal's *Elektra*," *Orbis Litterarum* 41 (1986): 312–31.

33. Hofmannsthal, "Elektra," in *Dramen II*, ed. Bernd Schoeller, *Gesammelte Werke in zehn Einzelbänden* (Frankfurt a.M.: Fischer Taschenbuch, 1979), 185–234, here 233. Hugo von Hofmannsthal, *Electra: A Tragedy in One Act*, in *Selected Plays and Libretti*, ed. Michael Hamburger, trans. Alfred Schwarz (New York: Pantheon Books, 1963), 1–77, here 77.

34. Hofmannsthal, "Elektra," 233–34; *Electra*, 77.

35. Brandstetter, "Der Traum vom anderen Tanz," 42. Alexandra Kolb also argues that Hofmannsthal, along with other early twentieth-century writers, imagined dance to be a specifically feminine medium. Kolb shows how male writers purposely ignored the intellectual dimensions of female dancers' work in order to further their conception of dance as "total" and "immediate."

Alexandra Kolb, *Performing Femininity: Dance and Literature in German Modernism* (Oxford: Peter Lang, 2009), 46.

36. David E. Wellbery, "Die Opfer-Vorstellung als Quelle der Faszination: Anmerkungen zum Chandos-Brief und zur frühen Poetik Hofmannsthals," *Hofmannsthal Jahrbuch* 11, ed. Gerhard Neumann, Ursual Renner, Günter Schnitzler, and Gotthart Wunberg (Freiburg i.Br.: Rombach, 2003), 281–310.

37. See Hofmannsthal, *Dramen IV*, 69–70.

38. Hugo von Hofmannsthal, "Amor und Psyche," *Dramen VI*, ed. Bernd Schoeller, *Gesammelte Werke in zehn Einzelbänden* (Frankfurt a.M.: Fischer Taschenbuch, 1979), 79–84, here 83.

39. Hugo von Hofmannsthal to Grete Wiesenthal, 1910, in *Grete Wiesenthal: Die Schönheit der Sprache des Körpers im Tanz*, ed. Leonhard M. Fiedler and Martin Lang (Salzburg: Residenz, 1985), 96–97, here 97.

40. *Grete Wiesenthal*, ed. Fiedler and Lang, 96.

41. *Grete Wiesenthal*, ed. Fiedler and Lang, 99.

42. This is from the essay "Augenblicke in Griechenland" (Moments in Greece), where Hofmannsthal distinguishes between disingenuous linguistic media (ancient literature and philosophy) and the nonarbitrary forms of material objects that take their shape from the female body. See Hofmannsthal, *Erzählungen*, 617–28.

43. On Hofmannsthal's acquaintance with Bachofen's work, see Peter Davies, *Myth, Matriarchy and Modernity: Johann Jakob Bachofen in German Culture, 1860–1945* (New York: de Gruyter, 2010), especially page 210.

44. Johann Jakob Bachofen, *Myth, Religion, and Mother Right: Selected Writings of J. J. Bachofen*, trans. Ralph Manheim (Princeton, NJ: Princeton University Press, 1967), 91.

45. Bachofen, *Mother Right*, 109–10.

46. See Sophie Salvo, "Father Is Always Uncertain: J. J. Bachofen and the Epistemology of Patriarchy," *Monatshefte* 116, no. 1 (Winter 2024): 44–65.

47. Hofmannsthal, "Elektra," 194; *Electra*, 17.

48. In the published book, Musil gave *Vereinigungen* the subtitle "zwei Erzählungen" (two narratives), while *Drei Frauen* is subtitled "Novellen" (novellas). Although Musil does refer to the stories of *Vereinigungen* as "Novellen" in his notebook entries, I will use the term *narrative* here to refer to "The Culmination of a Love" and "Die Versuchung der stillen Veronika" in keeping with the designation of Musil's published book.

49. Robert Musil, *Young Törless*, trans. Eithne Wilkins and Ernst Kaiser (London: Secker & Warburg, 1955), ix. Discussed in Patrizia McBride, *The Void of Ethics: Robert Musil and the Experience of Modernity* (Evanston, IL: Northwestern University Press, 2006), 27.

50. McBride, *The Void of Ethics*, 27.

51. McBride, *The Void of Ethics*, 28.

52. Robert Musil, "Ansätze zu einer neuen Ästhetik," *Gesammelte Werke in 9 Bänden*, ed. Adolf Frisé, vol. 8: *Essays und Reden* (Reinbek bei Hamburg: Rowohlt, 1981), 1137–54, here 1144; "Toward a New Aesthetic: Observations on a Dramaturgy of Film," in *Precision and Soul: Essays and Addresses*, ed. and trans. Burton Pike and David S. Luft (Chicago: University of Chicago Press, 1990), 193–207, here 199.

53. Nicola Gess, *Primitive Thinking: Figuring Alterity in German Modernity*, trans. Erik Butler and Susan L. Solomon (Berlin: de Gruyter, 2022), 248.

54. Robert Musil, *Tagebücher, Aphorismen, Essays und Reden* (Hamburg: Rowohlt, 1955), 188.

55. Musil, *Gesammelte Werke*, 2:1315.

56. Musil, *Gesammelte Werke*, 2:1314.

57. Robert Musil, "Die Vollendung der Liebe," *Gesammelte Werke*, 2:156–94, here 174; "The Culmination of Love," in *Intimate Ties: Two Novellas*, trans. Peter Wortsman (New York: Archipelago Books, 2019), 9–109, here 58.

58. The *Reallexikon der deutschen Literaturwissenschaft* programmatically describes the genre as: "Whereby what is narrated is to be understood as a temporally organized sequence of action, in which at least one character experiences a dynamic change of situation." Manfred Schmeling and Kerst Walstra, "Erzählung," in *Reallexikon der deutschen Literaturwissenschaft*, ed. Klaus Weimar, 3rd ed. (Berlin: de Gruyter, 2007), 1:517–22, here 517.

59. This is according to Jürgen Schröder, "Am Grenzwert der Sprache: Zu Robert Musils 'Vereinigungen,'" *Euphorion* 60 (1966): 311–34.

60. Musil, "Vollendung," 163–64; "Culmination of Love," 31–32.

61. Musil, "Vollendung," 184; "Culmination of Love," 83–84.

62. Musil, "Vollendung," 183; "Culmination of Love," 82, translation modified.

63. Musil, "Vollendung," 192; "Culmination of Love," 104–5.

64. Musil, "Vollendung," 193. Musil, "Culmination of Love," 109.

65. Musil, "Vollendung," 193. Musil, "Culmination of Love," 107.

66. Musil, Vollendung," 193; "Culmination of Love," 108–9, translation modified.

67. Although the relationship between Claudine and the undersecretary is often described as "sexual" in secondary scholarship, the text does not actually show them consummate their relationship—suggesting that this physical encounter, too, would be beyond its ability to represent. Instead, the consummation takes place in language, when Claudine switches from the formal "Sie" to the informal "du." Once this shift occurs, the text can end—"And then with a shudder she felt, despite all, her body fill up with lust. [. . .] And very vaguely, the way children picture God, that He's great, she had an intimation of her love." Musil, "Vollendung," 193; "Culmination of Love," 109.

68. Robert Musil, "Tonka," *Gesammelte Werke*, 2:270–306, here 288; "Tonka," in *Five Women*, trans. Eithne Wilkins and Ernst Kaiser (Boston: Verba Mundi, 1999), 69–122, here 95.

69. Musil, "Tonka," 283; *Five Women*, 88.

70. Musil, "Tonka," 289 and 296; *Five Women*, 97 and 108.

71. Johann Wolfgang von Goethe famously defined the novella as a "sich ereignete unerhörte Begebenheit" (an unheard-of event that has actually occurred). See Johann Peter Eckermann, "Donnerstag abend, den 29. Januar 1827," in *1823–1832*, vol. 1 of *Gespräche mit Goethe in den letzten Jahren seines Lebens* (Leipzig: Brockhaus, 1836), 315–22, here 319.

72. Musil, "Tonka," 276; *Five Women*, 78.

73. Musil, "Grigia," 234–52, here 238–39; *Five Women*, 15–41, here 20–22.

74. Musil, "Grigia," 245; *Five Women*, 30.

75. See Musil, "Grigia," 243–44; *Five Women*, 28–29.

76. Musil, "Grigia," 247; *Five Women*, 34.

77. Musil, "Grigia," 235; *Five Women*, 16.

78. "There is a time in life when everything perceptibly slows down, as though one's life were hesitating to go on or trying to change its course. It may be that at this time one is more liable to disaster." Musil, "Grigia," 234; *Five Women*, 15.

79. Musil, "Tonka," 270; *Five Women*, 69.

80. Musil, "Tonka," 270; *Five Women*, 69, translation modified.

81. Robert Musil, "[Form und Inhalt/Ohne Titel—um 1910]," in *Gesammelte Werke* 2:1299–1302, here 1301.

82. Walter Benjamin, "Das Gespräch," in *Gesammelte Schriften*, vol. 2-1, ed. Rolf Tiedemann and Hermann Schweppenhäuser (Frankfurt a.M.: Surhrkamp, 1977), 91–96, here 95; "The Metaphysics of Youth," in *Selected Writings*, 4 vols., ed. Marcus Bullock and Michael W. Jennings (Cambridge, MA: Harvard University Press, 1996), 1:6–17, here 9, translation modified.

83. Benjamin, "Das Gespräch," 95; "Metaphysics," 9.

84. Benjamin, "Das Gespräch," 95; "Metaphysics," 10, translation modified.

85. Walter Benjamin, "Über die Sprache überhaupt und über die Sprache des Menschen," in *Gesammelte Schriften*, vol. 2-1, 140–57, here 147; "On Language as Such and on the Language of Man," in *Selected Writings*, 1:62–74, here 67.

86. Benjamin is writing here about Richard Paget. See "Probleme der Sprachsoziologie," in *Gesammelte Schriften*, vol. 3, ed. Hella Tiedemann-Bartels (Frankfurt a.M.: Surhrkamp, 1991), 452–80, here 478; Benjamin, "Problems in the Sociology of Language," trans. Edmund Jephcott, in *Selected Writings*, 3:68–93, here 84.

87. Gess, *Primitive Thinking*, 303–57.

88. Eva Geulen, "Toward a Genealogy of Gender in Walter Benjamin's Writing," *German Quarterly* 69, no. 2 (Spring 1996): 161–80, here 170.

89. Weigel, *Body- and Image-Space*, 82.

90. Benjamin, "Das Gespräch," 92; "Metaphysics," 7.

91. This is a selection from Benjamin's June 23, 1913 letter to Herbert Belmore. See Walter Benjamin, *Early Writings, 1910–1917*, trans. Howard Eiland (Cambridge, MA: Harvard University Press, 2011), 158–9. Quoted in Weigel, *Body- and Image-Space*, 81.

Coda

1. Peter Eisenberg, "Debatte um den Gender-Stern: Finger weg vom generischen Maskulinum!" *Tagesspiegel*, August 8, 2018.

2. Peter Eisenberg, "Das Gendern gefährdet unser höchstes Kulturgut: Deutsch als einheitliche Sprache," *Berliner Zeitung*, August 2, 2022.

3. Jürgen Trabant, "Es knackt im Gebälk der Republik," *Die Welt*, May 17, 2021.

4. Trabant, "Es knackt im Gebälk der Republik."

5. Paul Warren documents the existence of "uptalk," or "rising intonation on declarative utterances," its association with the female sex, and its negative reception in the popular press in *Uptalk: The Phenomenon of Rising Intonation* (Cambridge: Cambridge University Press, 2016), xiii. Regarding "creaky voice," Lisa Davidson shows that although the phenomenon is commonly associated with women, there is in fact little difference in listeners' ability to hear creaky voice in male and female speakers. See Lisa Davidson, "The Effects of Pitch, Gender, and Prosodic Context on the Identification of Creaky Voice," *Phonetica* 76, no. 4 (2019): 235–62.

6. Elizabeth S. Corredor, "Unpacking 'Gender Ideology' and the Global Right's Antigender Countermovement," *Signs: Journal of Women in Culture and Society* 44, no. 3 (2019): 613–38. The transnational character of "antigenderism" has been the subject of significant scholarship in the past five years. See, for instance, *Anti-Genderismus in Europa: Allianzen von Rechtspopulismus und religiösem Fundamentalismus*, ed. Sonja A. Strube, Rita Perintfalvi, Raphaela Hemet, Miriam Metze, and Cicek Sahbaz (Bielefeld: Transcript, 2021).

7. Roger Cohen and Léontine Gallois, "In a Nonbinary Pronoun, France sees a U.S. Attack on the Republic," *New York Times*, November 28, 2021.

8. Karina Piser, "Aux Armes, Citoyen·nes!" *Foreign Policy*, July 4, 2021. On the mobilization of ideas about protecting children by right-wing antigender discourses, see Agnieszka Graff and Elzbieta Korolczuk, "Anxious Parents and Children in Danger: The Family as a Refuge from Neoliberalism," in *Anti-Gender Politics in the Populist Movement* (Abingdon: Routledge, 2021), 114–36.

9. Ana Lankes, "In Argentina, One of the World's First Bans on Gender-Neutral Language," *New York Times*, July 20, 2022.

10. Natalia Borza, "Why Shall I Call You *Ze*? Discourse Analysis of the Social Perception of Institutionally Introducing the Gender-Neutral Pronoun *Ze*," *Linguistik Online* 106, no. 1 (2021): 19–45.

11. Gal argues that we must view the "gender ideology" rhetoric of the far right not in terms of its propositionality, but of its pragmatics and metapragmatics: "What positions it occupies in what social fields and what opponents it indexes or evokes." Susan Gal, "Gender and the Discursive Authority of Far Right Politics," *Gender and Language* 15, no. 1 (2021): 96–103, here 100. Nicole Doerr makes a similar argument regarding images, showing how the AfD has appropriated images of "women's empowerment and LGBT rights discourse" even as it "maintain[s] its illiberal political agenda on gender and sexuality." See Nicole Doerr, "The Visual Politics of the Alternative for Germany (AfD): Anti-Islam, Ethno-Nationalism, and Gendered Images," *Social Sciences* 10, no. 1 (January 2021), https://doi.org/10.3390/socsci10010020.

12. "Gericht weist Klage gegen gendergerechte Sprache bei Audi ab," *Die Zeit*, July 29, 2022. Legally, the case was unsuccessful.

13. Wilhelm von Humboldt, "On Thinking and Speaking: Sixteen Theses on Language (1795/6)," trans. by Kurt Mueller-Vollmer in "Thinking and Speaking: Herder, Humboldt and Saussurean Semiotics," *Comparative Criticism* 11 (1989): 193–214.

14. Marcel Fürstenau, "Is Germany's Far-Right AfD a Threat to Democracy?" *Deutsche Welle*, January 2, 2024.

15. Seongcheol Kim, "The Populism of the Alternative for Germany (AfD): An Extended Essex School Perspective," *Palgrave Communications* 3, no. 5 (2017): 1–11, here 4.

16. Kim points out some exceptions to this: the AfD in Berlin, for example, has been relatively more open. See Kim, "Populism," 8.

17. Alternative für Deutschland, "Grundsatzprogramm für Deutschland," 2016, https://www.afd.de/grundsatzprogramm, 41; Alternative für Deutschland, "Manifesto for Germany," 2017, https://www.afd.de/wp-content/uploads/2017/04/2017-04-12_afd-grundsatzprogramm-englisch_web.pdf, 40.

18. "Programm," 47; "Manifesto," 46.

19. "Programm," 47; "Manifesto," 46.

20. Michael Townson discusses how, in the nineteenth century, "linguistic purism ceased to be regarded as a means to an end, but became an end in itself," a way to protect the "German 'Geist'" from foreign influence. Townson, *Mother-Tongue and Fatherland*, 98.

21. Claus Ahlzweig, *Muttersprache—Vaterland: Die deutsche Nation und ihre Sprache* (Wiesbaden: Westdeutscher, 1994).

22. Katarzyna Jastal, "Körper und Geschlecht der deutschen Sprache im sprachnationalen Diskurs des 19. Jahrhunderts," in *Im Clash der Identitäten: Nationalismen im literatur- und kulturgeschichtlichen Diskurs*, ed. Wolfgang Brylla and Cezary Lipiński (Göttingen: V & R Unipress, 2020), 93–104.

23. Friedrich Ludwig Jahn and Ernst Eiselen, *Die deutsche Turnkunst zur Einrichtung der Turnplätze* (Berlin, 1816), xxi–xxii.

24. "Programm," 47; "Manifesto," 46–47. I am using the translation "genderised" rather than "gendered" because this is the word the AfD uses in its own English translation.

25. For more on the *Gendersternchen* (gender star or asterisk), see Helga Kotthoff, "Gender-Sternchen, Binnen-I oder generisches Maskulinum, . . . (Akademische) Textstile der Personen-referenz als Registrierungen?" *Linguistik Online* 103, no. 3 (2020): 105–27. On the public recep-tion of the gender asterisk, see Vít Kolek, "Discourse of Non-Heteronormative Labelling in German-Language Press: The Case of Gendersternchen," *Slovenščina* 7, no. 2 (November 2019): 118–40. On gender in the German-language classroom, see Anगineh Djavadghazaryans, " 'Please Don't Gender Me!' Strategies for Inclusive Language Instruction in a Gender-Diverse Campus Community," in *Diversity and Decolonization in German Studies*, ed. Regine Criser and Ervin Malakaj (New York: Palgrave Macmillan, 2020), 269–87.

26. This practice originated with the German linguist Lann Hornscheidt and faced signifi-cant backlash in the German press. See Carrie Smith-Prei and Maria Stehle, *Awkward Politics: Technologies of Popfeminist Activism* (Montreal: McGill-Queen's University Press, 2016), 197–98.

27. The function of *xier* is outlined at https://www.annaheger.de. Nichole M. Neuman gives a helpful overview of some of the gender-neutral pronouns used by German speakers in "Illi Anna Heger's Grammatical Futurity," *Seminar* 56, no. 3–4 (2020): 302–21.

28. Hornscheidt's work is extensive. In addition to *feministische w_orte*, their publications include *Die sprachliche Benennung von Personen aus konstruktivistischer Sicht: Genderspezifi-zierung und ihre diskursive Verhandlung im heutigen Schwedisch* (Berlin: de Gruyter, 2006) and, with Ja'n Sammla, *Wie schreibe ich divers? Wie spreche ich gendergerecht? Ein Praxis-Handbuch zu Gender und Sprache* (Hiddensee: w_orten & meer, 2021).

29. Cabala de Sylvain and Carsten Balzer, "Die SYLVAIN-Konventionen—Versuch einer 'geschlechtergerechten' Grammatik-Transformation der deutschen Sprache," *Liminalis* 2 (2008): 40–53. See https://www.geschlechtsneutral.net for more on this and other gender-inclusive innovations.

30. "Programm," 47; "Manifesto," 48.

31. Historical Speeches TV, "Steffen Königer über Genderwahn (Original AfD-Video)," You-Tube video, January 16, 2018, https://www.youtube.com/watch?v=QTsbRIWbJfo.

32. "Programm," 55; "Manifesto," 54.

33. For more on this, see André Keil, " 'We need to rediscover our manliness . . .': The Lan-guage of Gender and Authenticity in German Right-Wing Populism," *Journal of Language and Politics* 19, no. 1 (2020): 107–24.

34. "Programm," 49; "Manifesto," 48.

35. "Programm," 55; "Manifesto," 54.

36. "Programm," 55; "Manifesto," 54. For the history of the term *gender ideology* and its ori-gins in communications from the Vatican, see Corredor, "Unpacking 'Gender Ideology.' " On the "paradox" of women leaders in a party that advocates for "traditional" gender roles, see Gabriele Dietze, "Rechtspopulismus und Geschlecht: Paradox und Leitmotiv," *Femina Politica* 27, no. 1 (2018): 34–46.

Bibliography

Aarsleff, Hans. "The Context and Sense of Humboldt's Statement That Language 'Ist kein Werk (Ergon), sondern eine Tätigkeit (Energeia).'" In *History of Linguistics 2002: Selected Papers from the Ninth International Conference on the History of the Language Sciences, 27–30 August 2002, São Paulo—Campinas*, edited by Eduardo Guimarães and Diana Luz Pessoa de Barros, 197–206. Amsterdam: John Benjamins, 2007.

Aarsleff, Hans. *From Locke to Saussure: Essays on Study of Language and Intellectual History*. Minneapolis: University of Minnesota Press, 1985.

Aarsleff, Hans. "The Rise and Decline of Adam and his *Ursprache* in Seventeenth-Century Thought." In *The Language of Adam/Die Sprache Adams*, edited by Allison P. Coudert, 277–95. Wiesbaden: Harrassowitz, 1999.

Abizadeh, Arash. "Was Fichte an Ethnic Nationalist? On Cultural Nationalism and Its Double." *History of Political Thought* 26, no. 2 (Summer 2005): 334–59.

Académie françoise. *Dictionnaire de l'Académie françoise dedié au Roy*. 2 vols. Paris: La veuve de Jean-Baptiste Coignard, 1694.

Adam, Lucien. "Du parler des hommes et des femmes dans la langue caraïbe." *Revue de linguistique et de philologie comparée* 12 (1879): 275–304.

Adelung, Johann Christoph. *Deutsche Sprachlehre*. Berlin: Christian Friedrich Voß und Sohn, 1781.

Adelung, Johann Christoph. "Von dem Geschlechte der Hauptwörter." In *Umständliches Lehrgebäude der Deutschen Sprache, zur Erläutertung der Deutschen Sprachlehre für Schulen*, vol. 1, 343–69. Leipzig: Breitkopf, 1782.

Adelung, Johann Christoph. "Von dem Geschlechte der Substantive." *Magazin für die deutsche Sprache* 1, no. 4 (1783): 3–20.

Ahlzweig, Claus. *Muttersprache—Vaterland: Die deutsche Nation und ihre Sprache*. Wiesbaden: Westdeutscher, 1994.

Ahmed, Sara. *Queer Phenomenology: Orientations, Objects, Others*. Durham, NC: Duke University Press, 2006.

Ahmed, Siraj. *Archaeology of Babel: The Colonial Foundations of the Humanities*. Stanford, CA: Stanford University Press, 2018.

Albertan-Coppola, Sylviane. "Les langues dans les *Moeurs des sauvages* de Lafitau." *Dix-Huitième Siècle* 22 (1990): 127–38.

Albisetti, James C. "German Influence on the Higher Education of American Women, 1865–1914." In *German Influences on Education in the United States to 1917*, edited by Henry Geitz, Jürgen Heideking, and Jurgen Herbst, 227–44. Cambridge: Cambridge University Press, 1995.

Albisetti, James C. *Schooling German Girls and Women: Secondary and Higher Education in the Nineteenth Century*. Princeton, NJ: Princeton University Press, 1988.

Allaire, Louis. "The Caribs of the Lesser Antilles." In *The Indigenous People of the Caribbean*, edited by Samuel M. Wilson, 177–85. Gainesville: University Press of Florida, 1997.

Allen, Ann Taylor. *Feminism and Motherhood in Germany, 1800–1914*. New Brunswick, NJ: Rutgers University Press, 1991.

Alternative for Germany [Alternative für Deutschland]. "Manifesto for Germany." Translated by Alternative für Deutschland. April 12, 2017. https://www.afd.de/wp-content/uploads/2017/04/2017-04-12_afd-grundsatzprogramm-englisch_web.pdf.

Alternative für Deutschland. "Grundsatzprogramm für Deutschland." Presented at the Bundesparteitag in Stuttgart, April 30–May 1, 2016. https://www.afd.de/grundsatzprogramm.

Althans, Birgit. *Der Klatsch, die Frauen und das Sprechen bei der Arbeit*. Frankfurt a.M.: Campus, 2000.

Anderson, Benedict. *Imagined Communities: Reflections on the Origin and Spread of Nationalism*. London: Verso, 2006.

Andreas-Salomé, Lou. "Der Mensch als Weib: Ein Bild im Umriss." In *Die Erotik: Vier Aufsätze*, edited by Ernst Pfeiffer, 7–44. Frankfurt a.M.: Ullstein, 1992.

Arens, Katherine. *Functionalism and Fin de Siècle: Fritz Mauthner's Critique of Language*. New York: Peter Lang, 1984.

Aristotle. *Generation of Animals*. Translated by A. L. Peck. Loeb Classical Library 366. Cambridge, MA: Harvard University Press, 1942.

Aristotle. *History of Animals: Volume 3, Books 7–10*. Edited and translated by D. M. Balme. Loeb Classical Library 439. Cambridge, MA: Harvard University Press, 1991.

Aristotle. *Minor Works*. Translated by W. S. Hett. Loeb Classical Library 307. Cambridge, MA: Harvard University Press, 1936.

Aristotle. *Problems*. Translated by E. S. Forster. In *Complete Works of Aristotle: The Revised Oxford Translation*, edited by Jonathan Barnes, vol. 2. Princeton, NJ: Princeton University Press, 1984.

Arnauld, Antoine, and Claude Lancelot. *General and Rational Grammar: The Port-Royal Grammar*. Edited and translated by Jacques Rieux and Bernard E. Rollin. The Hague: Mouton, 1975.

Arnauld, Antoine, and Claude Lancelot. *Grammaire générale et raisonnée de Port-Royal*. Paris: Perlet, 1803.

Arvidsson, Stefan. *Aryan Idols: Indo-European Mythology as Ideology and Science*. Translated by Sonia Wichmann. Chicago: University of Chicago Press, 2006.

Auroux, Sylvain. "The First Uses of the French Word 'Linguistique' (1812–1880)." In *Papers in the History of Linguistics: Proceedings of the Third International Conference on the History of the Language Sciences (ICHoLS III), Princeton, 19–23 August 1984*, edited by Hans Aarsleff, L. G. Kelly, and Hans-Josef Niederehe, 447–59. Amsterdam: John Benjamins, 1987.

Auroux, Sylvain, E. F. K. Koerner, Hans-Josef Niederehe, and Kees Versteegh, eds. *History of the Language Sciences / Geschichte der Sprachwissenschaften / Histoire des Sciences du langage: An International Handbook on the Evolution of the Study of Language from the Beginnings to the Present*. 3 vols. Berlin: de Gruyter Mouton, 2000–06.

Ayres-Bennett, Wendy, and Helena Sanson, eds. *Women in the History of Linguistics*. Oxford: Oxford University Press, 2021.

Bachofen, Johann Jakob. *Myth, Religion, and Mother Right: Selected Writings of J. J. Bachofen*. Translated by Ralph Manheim. Princeton, NJ: Princeton University Press, 1967.

Bailey, Nathan. *An Universal Etymological English Dictionary*. 20th ed. London: T. Osborne, 1763.

Balibar, Étienne. "Fichte and the Internal Border: On *Addresses to the German Nation*." In *Masses, Classes, Ideas: Studies on Politics and Philosophy Before and After Marx*, 61–84. New York: Routledge, 1994.

Bank, Andrew. "Evolution and Racial Theory: The Hidden Side of Wilhelm Bleek." *South African History Journal* 43 (November 2000): 163–78.

Bardsley, Sandy. *Venomous Tongues: Speech and Gender in Late Medieval England*. Philadelphia: University of Pennsylvania Press, 2006.

Baron, Dennis. *Grammar and Gender*. New Haven, CT: Yale University Press, 1986.

Baron, Dennis. *What's Your Pronoun? Beyond He and She*. New York: Liveright, 2020.

Beauzée, Nicolas, and Jacques-Philippe-Augustin Douchet. "Genre (Grammaire)." In *Encyclopédie, ou dictionnaire raisonné des sciences, des arts et des métiers, etc.*, edited by Denis Diderot and Jean le Rond d'Alembert. University of Chicago: ARTFL Encyclopédie Project, 2016, edited by Robert Morrissey and Glenn Roe. http://encyclopedie.uchicago.edu.

Benecke, Georg Friedrich. Review of *Deutsche Grammatik* by Jacob Grimm. *Göttingische gelehrte Anzeigen* 201 (December 19, 1822): 2001–8.

Benes, Tuska. *In Babel's Shadow: Language, Philology, and the Nation in Nineteenth-Century Germany*. Detroit: Wayne State University Press, 2008.

Benfey, Theodor. *Geschichte der Sprachwissenschaft und der orientalischen Philologie in Deutschland seit dem Anfange des 19. Jahrhunderts mit einem Rückblick auf die früheren Zeiten*. Munich: J. G. Cotta, 1869.

Benjamin, Walter. *Early Writings, 1910–1917*. Translated by Howard Eiland. Cambridge, MA: Harvard University Press, 2011.

Benjamin, Walter. *Gesammelte Schriften*. Edited by Rolf Tiedemann and Hermann Schweppenhäuser. 7 vols. Frankfurt a.M.: Suhrkamp, 1991.

Benjamin, Walter. *Selected Writings*. Edited by Marcus Bullock and Michael W. Jennings. 4 vols. Cambridge, MA: Belknap Press of Harvard University Press, 1996–2003.

Ben-Zvi, Linda. "Samuel Beckett, Fritz Mauthner, and the Limits of Language." *PMLA* 95, no. 2 (March 1980): 183–200.

Bernasconi, Robert. "Who Invented the Concept of Race? Kant's Role in the Enlightenment Construction of Race." In *Race*, edited by Robert Bernasconi, 11–36. Malden, MA: Blackwell, 2001.

Bernhardi, August Ferdinand. *Anfangsgründe der Sprachwissenschaft*. Berlin: Heinrich Frölich, 1805.

Bernhardi, August Ferdinand. *Sprachlehre*. 2 vols. Berlin: Heinrich Frölich, 1801–3.

Bleek, Wilhelm. *A Comparative Grammar of South African Languages*. London: Trübner, 1862.

Bleek, Wilhelm. *On the Origin of Language*. Edited by Ernst Haeckel. Translated by Thomas Davidson. New York: L. W. Schmidt, 1869.

Bleek, Wilhelm. *Über den Ursprung der Sprache*. Cape Town: Van de Sandt de Villiers, 1867.

Boetcher Joeres, Ruth-Ellen. *Respectability and Deviance: Nineteenth-Century German Women Writers and the Ambiguity of Representation*. Chicago: University of Chicago Press, 1998.

Bonald, Louis de. "De l'origine du langage." In *Récherches philosophiques sur les premiers objets des connaissances morales*, 61–117. Paris: A. Le Clere, 1882.

Borza, Natalia. "Why Shall I Call You *Ze*? Discourse Analysis of the Social Perception of Institutionally Introducing the Gender-Neutral Pronoun *Ze*." *Linguistik Online* 106, no. 1 (2021): 19–45.

Bottigheimer, Ruth B. *Grimm's Bad Girls and Bold Boys: The Moral and Social Vision of the Tales*. New Haven, CT: Yale University Press, 1987.

Boucher, Philip P. *Cannibal Encounters: Europeans and Island Caribs, 1492–1763*. Baltimore: Johns Hopkins University Press, 1992.

Boyd, Zelda. "'The Grammarian's Funeral' and the Erotics of Grammar." *Browning Institute Studies* 16 (1988): 1–14.

Brandstetter, Gabriele. "Der Traum vom anderen Tanz: Hofmannsthals Ästhetik des Schöpferischen im Dialog 'Furcht.'" In *Hugo von Hofmannsthal: Neue Wege der Forschung*, edited by Elsbeth Dangel-Pelloquin, 41–61. Darmstadt: Wissenschaftliche Buchgesellschaft, 2007.

Braun, Christina von. "Männliche und weibliche Form in Natur und Kultur in der Wissenschaft." Paper presented as part of the "Humboldt Gespräche" at the Conference for the Bundeszentrale für politische Bildung, Berlin, Germany, June 7–8, 2007.

Bredeck, Elizabeth. *Metaphors of Knowledge: Language and Thought in Mauthner's Critique*. Detroit: Wayne State University Press, 1992.

Breton, Raymond. *Dictionnaire françois-caraïbe*. Auxerre: Gilles Bouquet, 1666.

Brill, Willem Gerard. *Hollandsche Spraakleer*. Leiden: Luchtmans, 1846.

Brugmann, Karl. "Das Nominalgeschlecht in den indogermanischen Sprachen." *Internationale Zeitschrift für allgemeine Sprachwissenschaft* 4 (1889): 100–109.

Brugmann, Karl. *On the Nature and Origin of the Noun Genders in the Indo-European Languages*. Translated by Edmund Y. Robbins. New York: Charles Scribner's Sons, 1897.

Brugmann, Karl. "Zur Frage der Entstehung des grammatischen Geschlechts." *Beiträge zur Geschichte der deutschen Sprache und Literatur* 15 (1891): 523–31.

Buffon, Comte de [Georges-Louis Leclerc]. *Histoire des Animaux quadrupèdes*. Vol. 1 of *Œuvres completes*. Paris: De l'Imprimerie Royal, 1775.

Butler, Judith. *Bodies that Matter: On the Discursive Limits of "Sex."* London: Routledge, 1993.

Butterworth, Emily. *The Unbridled Tongue: Babble and Gossip in Renaissance France*. Oxford: Oxford University Press, 2016.

Cameron, Deborah. *Feminism and Linguistic Theory*. London: Palgrave Macmillan, 1985.

Cameron, Deborah. *The Myth of Mars and Venus: Do Men and Women Really Speak Different Languages?* Oxford: Oxford University Press, 2008.

Campbell, Mary Baine. *Wonder and Science: Imagining Worlds in Early Modern Europe*. Ithaca, NY: Cornell University Press, 1999.

Campe, Joachim Heinrich. *Väterlicher Rath für meine Tochter*. Frankfurt a.M.: Cotta, 1789.

Carson, Anne. "The Gender of Sound." In *Glass, Irony, and God*, 119–39. New York: New Directions, 1995.

Cassirer, Ernst. "The Form of the Concept in Mythical Thinking (1922)." In *The Warburg Years (1919–1933): Essays on Language, Art, Myth, and Technology*, translated by S. G. Lofts, 1–71. New Haven, CT: Yale University Press, 2013.

Cavallaro, Dani. *French Feminist Theory: An Introduction*. London: Continuum, 2003.

Chamberlain, Alexander F. "Women's Languages." *American Anthropologist* 14, no. 3 (July–September 1912): 579–81.

Chomsky, Noam. *Cartesian Linguistics: A Chapter in the History of Rationalist Thought*. New York: Harper & Row, 1966.

Cicero. *De Oratore*. Translated by William Guthrie. Oxford: Henry Slatter, 1840.

Cixous, Hélène. "The Laugh of the Medusa." Translated by Keith Cohen and Paula Cohen. *Signs* 1, no. 4 (Summer 1976): 875–93.

Cixous, Hélène. "Le Rire de la Méduse." *L'Arc* 61 (May 1975): 39–54.

Cohen, Roger, and Léontine Gallois. "In a Nonbinary Pronoun, France Sees a U.S. Attack on the Republic." *New York Times*, November 28, 2021.

Condillac, Étienne Bonnot de. *Essai sur l'origine des connaissances humaines*. Vol. 1 of *Ouevres complètes de Condillac*. Paris: Ch. Houel, 1798.

Condillac, Étienne Bonnot de. *An Essay on the Origin of Human Knowledge*. Translated by Thomas Nugent. London: J. Nourse, 1756.

Cooper, Vincent O. "Language and Gender among the Kalinago of Fifteenth-century St. Croix." In *The Indigenous People of the Caribbean*, edited by Samuel M. Wilson, 186–96. Gainesville: University Press of Florida, 1997.

Copineau, Abbé. *Essai synthétique sur l'origine et la formation des langues*. Paris: Ruault, 1774.

Corbeill, Anthony. *Sexing the World: Grammatical Gender and Biological Sex in Ancient Rome*. Princeton, NJ: Princeton University Press, 2015.

Corkhill, Alan. "Female Language Theory in the Age of Goethe: Three Case Studies." *Modern Language Review* 94, no. 4 (October 1999): 1041–53.

Cormican, Muriel. *Women in the Works of Lou Andreas-Salomé: Negotiating Identity*. Rochester, NY: Camden House, 2009.

Corredor, Elizabeth S. "Unpacking 'Gender Ideology' and the Global Right's Antigender Countermovement." *Signs: Journal of Women in Culture and Society* 44, no. 3 (2019): 613–38.

Court de Gébelin, Antoine. *Monde primitif, analysé et comparé avec le monde moderne, considéré dans l'histoire naturelle de la parole*. 2 vols. Paris: Boudet, 1774.

Czermak, Johann Nepomuk. *Gesammelte Schriften*. 2 vols. Leipzig: Wilhelm Engelmann, 1879.

Darwin, Charles. *The Descent of Man, and Selection in Relation to Sex*. 2 vols. London: John Murray, 1871.

Das häusliche Glück: Vollständiger Haushaltungsunterricht nebst Anleitung zum Kochen für Arbeiterfrauen. M. Gladbach: A. Riffarth, 1905.

Daub, Adrian. *Uncivil Unions: The Metaphysics of Marriage in German Idealism and Romanticism*. Chicago: University of Chicago Press, 2022.

Davidson, Lisa. "The Effects of Pitch, Gender, and Prosodic Context on the Identification of Creaky Voice." *Phonetica* 76, no. 4 (2019): 235–62.

Davies, Peter. *Myth, Matriarchy and Modernity: Johann Jakob Bachofen in German Culture, 1860–1945*. New York: de Gruyter, 2010.

Delaunay, Gaëtan. *Études de biologie comparée basées sur l'évolution organique*. 2 vols. Paris: V. Adrien Delahaye, 1878–79.

Delbrück, Berthold. *Einleitung in das Sprachstudium: Ein Beitrag zur Geschichte und Methodik der vergleichenden Sprachforschung*. Leipzig: Breitkopf & Härtel, 1880.

DeLeonibus, Gaetano. "Raymond Breton's 1665 'Dictionaire caraïbe-françois.'" *French Review* 80, no. 5 (April 2007): 1044–55.

Diamond, Cora. "The *Tractatus* and the Limits of Sense." In *The Oxford Handbook of Wittgenstein*, edited by Oskari Kuusela and Marie McGinn, 240–75. Oxford: Oxford University Press, 2011.

Dietrick, Linda. "Herder as a Mentor of Literary Women." In *Herder Jahrbuch 14*, edited by Rainer Godel and Johannes Schmidt, 119–43. Heidelberg: Synchron, 2018.

Dietze, Gabriele. "Rechtspopulismus und Geschlecht: Paradox und Leitmotiv." *Femina Politica* 27, no. 1 (2018): 34–46.

Die Zeit. "Gericht weist Klage gegen gendergerechte Sprache bei Audi ab." July 29, 2022.

Djavadghazaryans, Angineh. " 'Please Don't Gender Me!' Strategies for Inclusive Language Instruction in a-Gender-Diverse Campus Community." In *Diversity and Decolonization in German Studies*, edited by Regine Criser and Ervin Malakaj, 269–87. New York: Palgrave Macmillan, 2020.

Doerr, Nicole. "The Visual Politics of the Alternative for Germany (AfD): Anti-Islam, Ethno-Nationalism, and Gendered Images." *Social Sciences* 10, no. 1 (January 2021). https://doi.org/10.3390/socsci10010020.

Domsch, Sebastian. "Language and the Edges of Humanity: Orang-Utans and Wild Girls in Monboddo and Peacock." *Zeitschrift für Anglistik und Amerikanistik* 56, no. 1 (January 2008): 1–11.

DuPuis, Mathias. *Relation de l'establissement d'une colonie française dans la Gardeloupe, isle de l'Amérique, et des moeurs des sauvages.* Basse-Terre: Société d'Histoire de la Guadeloupe, 1972.

Du Tertre, Jean-Baptiste. *Histoire générale des Antilles habitées par les François.* 4 vols. Paris: Thomas Iolly, 1667.

Ebbersmeyer, Sabrina. "From a 'Memorable Place' to 'Drops in the Ocean': On the Marginalization of Women Philosophers in German Historiography of Philosophy." *British Journal for the History of Philosophy* 28, no. 3 (May 2020): 442–62.

Eckermann, Johann Peter. "Donnerstag abend, den 29. Januar 1827." In *1823–1832*, 315–22. Vol. 1 of *Gespräche mit Goethe in den letzten Jahren seines Lebens.* Leipzig: Brockhaus, 1836.

Eckert, Penelope and Sally McConnell-Ginet. "Being Assertive . . . Or Not." In *Language and Gender*, 141–63. Cambridge: Cambridge University Press, 2013.

Eckert, Penelope, and Sally McConnell-Ginet. "Think Practically and Look Locally: Language and Gender as Community-Based Practice." *Annual Review of Anthropology* 21 (1992): 461–90.

Eisenberg, Peter. "Das Gendern gefährdet unser höchstes Kulturgut: Deutsch als einheitliche Sprache." *Berliner Zeitung*, August 2, 2022.

Eisenberg, Peter. "Debatte um den Gender-Stern: Finger weg vom generischen Maskulinum!" *Tagesspiegel*, August 8, 2018.

Elffers, Els. "The Rise of General Linguistics as an Academic Discipline: Georg von der Gabelentz as a Co-Founder." In *From Early Modern to Modern Disciplines*, 55–70. Vol. 2 of *The Making of the Humanities*, edited by Rens Bod, Jaap Maat, and Thijs Weststeijn. Amsterdam: Amsterdam University Press, 2012.

Engelstein, Stefani. "The Allure of Wholeness: The Eighteenth-Century Organism and the Same-Sex Marriage Debate." *Critical Inquiry* 39, no. 4 (Summer 2013): 754–76.

Engelstein, Stefani. *Sibling Action: The Genealogical Structure of Modernity.* New York: Columbia University Press, 2017.

Errington, Joseph. *Linguistics in a Colonial World: A Story of Language, Meaning, and Power.* Malden, MA: Blackwell, 2008.

Fausto-Sterling, Anne. *Sexing the Body: Gender Politics and the Construction of Sexuality.* New York: Basic Books, 2000.

Fiala, Andrew. "Fichte and the *Ursprache*." In *After Jena: New Essays on Fichte's Later Philosophy*, edited by Daniel Breazeale and Tom Rockmore, 183–97. Evanston, IL: Northwestern University Press, 2008.

Fichte, Johann Gottlieb. *Addresses to the German Nation*. Edited by Gregory Moore. Cambridge: Cambridge University Press, 2008.

Fichte, Johann Gottlieb. *Foundations of Natural Right*. Edited by Frederick Neuhouser. Translated by Michael Baur. Cambridge: Cambridge University Press, 2000.

Fichte, Johann Gottlieb. *Gesamtausgabe der Bayerischen Akademie der Wissenschaften*. Edited by Erich Fuchs, Hans Gliwitzky, Reinhard Lauth, and Peter K. Schneider. 42 vols. Stuttgart: Friedrich Frommann, 1962–2012.

Fichte, Johann Gottlieb. "On the Linguistic Capacity and the Origin of Language." Translated by Jere Paul Surber. In Jere Paul Surber, *Language and German Idealism: Fichte's Linguistic Philosophy*, 117–45. Atlantic Highlands, NJ: Humanities Press, 1996.

Fichte, Johann Gottlieb. "Some Lectures Concerning the Scholar's Vocation." In *Early Philosophical Writings*, edited and translated by Daniel Breazeale, 144–84. Ithaca, NY: Cornell University Press, 1988.

Fichte, Johann Gottlieb. *The System of Ethics: According to the Principles of the Wissenschaftslehre*. Edited and translated by Daniel Breazeale and Günter Zöller. Cambridge: Cambridge University Press, 2005.

Fiedler, Leonhard M., and Martin Lang, eds. *Grete Wiesenthal: Die Schönheit der Sprache des Körpers im Tanz*. Salzburg: Residenz, 1985.

Filmer, Robert. *Patriarcha, or the Natural Power of Kings*. London: R. Chiswell, 1680.

Fleischer, Mary. "Hugo von Hofmannsthal and Grete Wiesenthal." In *Embodied Texts: Symbolist Playwright-Dancer Collaborations*, 93–148. Amsterdam: Rodopi, 2007.

Fögen, Thorsten. "Gender-Specific Communication in Graeco-Roman Antiquity." *Historiographia Linguistica* 32, no. 2–3 (2004): 199–276.

Formey, Jean Henri Samuel. "Réunion des principaux moyens employés pour découvrir l'origine du langage, des idées, et des connais." In *Anti-Emile*, 211–30. Berlin: Joachim Pauli, 1763.

Formigari, Lia. "Idealism and Idealistic Trends in Linguistics and in the Philosophy of Language." In *Der epistemologische Kontext neuzeitlicher Sprach- und Grammatiktheorien*. Vol. 1 of *Sprachtheorien der Neuzeit*, edited by Peter Schmitter, 230–53. Tübingen: Gunter Narr, 1999.

Forster, Michael N. *After Herder: Philosophy of Language in the German Tradition*. Oxford: Oxford University Press, 2010.

Forster, Michael N. "Gods, Animals, and Artists: Some Problem Cases in Herder's Philosophy of Language." In *After Herder: Philosophy of Language in the German Tradition*, 91–130. Oxford: Oxford University Press, 2010.

Foucault, Michel. *The Order of Things: An Archaeology of the Human Sciences*. New York: Vintage Books, 2012.

Fox, Jennifer. "The Creator Gods: Romantic Nationalism and the En-genderment of Women in Folklore." *Journal of American Folklore* 100, no. 398 (October–December 1987): 563–72.

Frain du Tremblay, Jean. *Traité des langues: Où l'on donne des principes et des règles pour juger du mérite et de l'excellence de chaque langue et en particulier de la langue françoise*. Paris: Jean-Baptiste Delespine, 1703.

Frain du Tremblay, Jean. *A Treatise of Languages: Wherein Are Laid Down the General Principles of Each, with Proper Rules to Judge of Their Respective Merits and Excellence, and More Particularly, of the French and English*. Translated by M. Halpenn. London: D. Leach, 1725.

Franklin, Caroline. "'Quiet Cruising o'er the Ocean Woman': Byron's 'Don Juan' and the Woman Question." *Studies in Romanticism* 29, no. 4 (Winter 1990): 603–31.

Frazer, James G. "A Suggestion as to the Origin of Gender in Language." *Fortnightly Review* 73 (January 1900): 79–90.

Friedrichsmeyer, Sarah. *The Androgyne in Early German Romanticism: Friedrich Schlegel, Novalis and the Metaphysics of Love.* Bern: Peter Lang, 1983.

Frischmann, Bärbel. "Fichte's Theory of Gender Relations in his *Foundations of Natural Right.*" In *Rights, Bodies and Recognition: New Essays on Fichte's* Foundations of Natural Right, edited by Tom Rockmore and Daniel Breazeale, 152–65. Burlington, VT: Ashgate, 2006.

Fürstenau, Marcel. "Is Germany's Far-Right AfD a Threat to Democracy?" *Deutsche Welle,* January 2, 2024.

Furetière, Antoine. *Dictionaire universel.* 3 vols. The Hague: Arnoud et Reinier Leers, 1690.

Gabelentz, Georg von der. "Begriff der menschlichen Sprache." In *Die Sprachwissenschaft: Ihre Aufgaben, Methoden, und bisherigen Ergebnisse,* edited by Manfred Ringmacher and James McElvenny, 2–5. Berlin: Language Science Press, 2016.

Gabelentz, Georg von der. *Die Sprachwissenschaft: Ihre Aufgaben, Methoden und bisherigen Ergebnisse,* edited by Manfred Ringmacher and James McElvenny. Berlin: Language Science Press, 2016.

Gal, Susan. "Gender and the Discursive Authority of Far Right Politics." *Gender and Language* 15, no. 1 (2021): 96–103.

Gardt, Andreas. *Geschichte der Sprachwissenschaft in Deutschland.* New York: de Gruyter, 1999.

George, Alys X. *The Naked Truth: Viennese Modernism and the Body.* Chicago: University of Chicago Press, 2020.

Gerhard, Ute. *Frauenbewegung und Feminismus: Eine Geschichte seit 1789.* Munich: C. H. Beck, 2009.

Gess, Nicola. *Primitive Thinking: Figuring Alterity in German Modernity.* Translated by Erik Butler and Susan L. Solomon. Berlin: de Gruyter, 2022.

Geulen, Eva. "Toward a Genealogy of Gender in Walter Benjamin's Writing." *German Quarterly* 69, no. 2 (Spring 1996): 161–80.

Gilbert, Sandra M., and Susan Gubar. *The Madwoman in the Attic: The Woman Writer and the Nineteenth-Century Literary Imagination.* New Haven, CT: Yale University Press, 1979.

Gilmour, Rachael. "From Languages to Language: The Comparative Philologist in South Africa." In *Grammars of Colonialism: Representing Languages in Colonial South Africa,* 169–91. New York: Palgrave Macmillan, 2007.

Girard, Gabriel. *Les vrais principes de la langue Françoise: ou, La parole réduite en méthode, conformément aux lois de l'usage, en seize discours.* 2 vols. Paris: Le Breton, 1747.

Goldstein, Amanda Jo. *Sweet Science: Romantic Materialism and the New Logics of Life.* Chicago: University of Chicago Press, 2017.

Golumbia, David. "The Deconstruction of Philology." *boundary 2* 48, no. 1 (2021): 17–34.

Goodman, Katherine R., and Edith Waldstein, eds. *In the Shadow of Olympus: German Women Writers around 1800.* Albany: State University of New York, 1992.

Gottsched, Johann Christoph. *Grundlegung einer deutschen Sprachkunst.* Leipzig: Bernhard Christoph Breitkopf, 1748.

Graff, Agnieszka, and Elżbieta Korolczuk. "Anxious Parents and Children in Danger: The Family as a Refuge from Neoliberalism." In *Anti-Gender Politics in the Populist Movement,* 114–36. Abingdon: Routledge, 2021.

Grant, Ruth W. "John Locke on Women and the Family." In John Locke, *Two Treatises of Govern-ment* and *A Letter Concerning Toleration*, edited by Ian Shapiro, with essays by John Dunn, Ruth W. Grant, and Ian Shapiro, 286–308. New Haven, CT: Yale University Press, 2003.

Grimm, Jacob. *Deutsche Grammatik*. Vol. 3. Göttingen: Dieterich, 1831.

Grimm, Jacob. *Deutsche Grammatik*. Edited by Wilhelm Scherer, Gustav Roethe, and Edward Schröder. 4 vols. Gütersloh: C. Bertelsmann, 1870–98.

Grimm, Jacob. *Kleinere Schriften*. Edited by Karl Müllenhoff and Eduard Ippel. Berlin: Ferd. Dümmler, 1865–90.

Grimm, Jacob. *Über den Ursprung der Sprache*. Berlin: Ferd. Dümmler, 1866.

Grimm, Jacob, and Wilhelm Grimm. *Deutsches Wörterbuch von Jacob und Wilhelm Grimm*. 16 vols. Leipzig: S. Hirzel, 1854–1960.

Grimm, Jacob, and Wilhelm Grimm. *The Original Folk and Fairy Tales of the Brothers Grimm: The Complete First Edition*. Translated and edited by Jack Zipes. Princeton, NJ: Princeton University Press, 2015.

Grosz, Elizabeth. *Sexual Subversions: Three French Feminists*. Sydney: Allen & Unwin, 1989.

Habermas, Jürgen. *Der philosophische Diskurs der Moderne: Zwölf Vorlesungen*. Frankfurt a.M.: Suhrkamp, 1985.

Hacking, Ian. "Night Thoughts on Philology." In *Historical Ontology*, 140–51. Cambridge, MA: Harvard University Press, 2002.

Hamburger, Michael. "Art as Second Nature: The Figures of the Actor and the Dancer in the Works of Hugo von Hofmannsthal." In *Romantic Mythologies*, edited by Ian Fletcher, 225–41. London: Routledge, 1967.

Haraway, Donna. *Primate Visions: Gender, Race, and Nature in the World of Modern Science*. New York: Routledge, 1989.

Harpham, Geoffrey Galt. "Roots, Races, and the Return to Philology." *Representations* 106, no. 1 (Spring 2009): 34–62.

Hartley, Florence. *The Ladies' Book of Etiquette and Manual of Politeness: A Complete Hand Book for the Use of the Lady in Polite Society*. Boston: J. S. Locke, 1876.

Hartman, Saidiya. "Venus in Two Acts." *Small Axe* 12, no. 2 (June 2008): 1–14.

Healy, George R. "The French Jesuits and the Idea of the Noble Savage." *William and Mary Quarterly* 15, no. 2 (April 1958): 143–67.

Heintel, E. "Besonnenheit." In *Historisches Wörterbuch der Philosophie*, edited by Joachim Ritter, Karlfried Gründer, and Gottfried Gabriel, vol. 1, 848–49. Basel: Schwabe, 1971.

Helmstetter, Rudolf. "Entwendet: Hofmannsthals *Chandos-Brief*, die Rezeptionsgeschichte und die Sprachkrise." *Deutsche Vierteljahrsschrift für Literaturwissenschaft und Geistesgeschichte* 77, no. 3 (2003): 446–80.

Hendricks, Christina, and Kelly Oliver, eds. *Language and Liberation: Feminism, Philosophy, and Language*. Albany: State University of New York Press, 1999.

Herder, Johann Gottfried. "Gedichte von Sophie Mereau: Erstes Bändchen: Berlin 1800." In *Sämmtliche Werke: Zur Philosophie und Geschichte*, vol. 21, 392–98. Stuttgart: Cotta, 1830.

Herder, Johann Gottfried. *Herders Briefwechsel mit Caroline Flachsland*. Edited by Hans Schauer. 2 vols. Weimar: Goethe-Gesellschaft, 1926.

Herder, Johann Gottfried. *Outlines of a Philosophy of the History of Man*. Translated by T. Churchill. 2nd ed. 2 vols. London: Luke Hansard, 1803.

Herder, Johann Gottfried. "Paramythia: From the German of Herder." *Belfast Monthly Magazine* 2, no. 9 (April 30, 1809): 262–66.

Herder, Johann Gottfried. *Philosophical Writings.* Edited and translated by Michael N. Forster. Cambridge: Cambridge University Press, 2002.

Herder, Johann Gottfried. *Selected Writings on Aesthetics.* Edited and translated by Gregory Moore. Princeton, NJ: Princeton University Press, 2006.

Herder, Johann Gottfried. *Werke in zehn Bänden.* Edited by Ulrich Gaier. 10 vols. Frankfurt a.M.: Suhrkamp, 1985–2000.

Herodotus. *The Histories.* Translated by Robin Waterfield. Oxford: Oxford University Press, 1998.

Herodotus. *Les neuf livres des histoires d'Hérodote, prince et premier des Historiographes Grecz, intitulez du nom des Muses.* Translated by Pierre Saliat. Paris: Jean de Roigny, 1556.

Historical Speeches TV. "Steffen Königer über Genderwahn (Original AfD-Video)." YouTube video. January 16, 2018. https://www.youtube.com/watch?v=QTsbRIWbJfo.

Hockett, Charles Francis. *A Course in Modern Linguistics.* New York: Macmillan, 1958.

Hoenigswald, Henry M. "Nineteenth-Century Linguistics on Itself." In *Studies in the History of Western Linguistics,* edited by Theodora Bynon and F. R. Palmer, 172–88. Cambridge: Cambridge University Press, 1986.

Hoffrath, Christiane. *Bücherspuren: Das Schicksal von Elise und Helene Richter und ihrer Bibliothek im "Dritten Reich."* Cologne: Böhlau, 2009.

Hofmannsthal, Hugo von. *Electra: A Tragedy in One Act.* In *Selected Plays and Libretti,* edited by Michael Hamburger, translated by Alfred Schwarz, 1–77. New York: Pantheon Books, 1963.

Hofmannsthal, Hugo von. *Gesammelte Werke in zehn Einzelbänden.* Edited by Bernd Schoeller and Rudolf Hirsch. 10 vols. Frankfurt a.M.: Fischer Taschenbuch, 1979–80.

Hofmannsthal, Hugo von. *Selected Prose.* Translated by Mary Hottinger, James Stern, and Tania Stern. New York: Pantheon Books, 1952.

Holenstein, Pia, and Norbert Schindler. "Geschwätzgeschichte(n): Ein kulturhistorisches Plädoyer für die Rehabilitierung der unkontrollierten Rede." In *Dynamik der Tradition,* edited by Richard van Dülmen, 41–108. Frankfurt a.M.: Fischer, 1992.

Holland, Jocelyn. *German Romanticism and Science: The Procreative Poetics of Goethe, Novalis, and Ritter.* New York: Routledge, 2009.

Holzhey, Christoph F. E. "On the Emergence of Sexual Difference in the 18th Century: Economies of Pleasure in Herder's *Liebe und Selbstheit.*" *German Quarterly* 79, no. 1 (Winter 2006): 1–27.

Honegger, Claudia. *Die Ordnung der Geschlechter: Die Wissenschaften vom Menschen und das Weib, 1750–1850.* Frankfurt a.M.: Campus, 1991.

Hornscheidt, Lann. *Die sprachliche Benennung von Personen aus konstruktivistischer Sicht: Genderspezifizierung und ihre diskursive Verhandlung im heutigen Schwedisch.* Berlin: de Gruyter, 2006.

Hornscheidt, Lann. *feministische w_orte: ein lern-, denk- und handlungsbuch zu sprache und diskriminierung, gender studies und feministischer linguistic.* Frankfurt a.M.: Brandes & Apsel, 2012.

Hornscheidt, Lann, and Ja'n Sammla. *Wie schreibe ich divers? Wie spreche ich gendergerecht? Ein Praxis-Handbuch zu Gender und Sprache.* Hiddensee: w_orten & meer, 2021.

Howatt, A. P. R., and Richard Smith. "The History of Teaching English as a Foreign Language, from a British and European Perspective." *Language and History* 57, no. 1 (May 2014), 75–95.

Hull, Isabel V. *Sexuality, State, and Civil Society in Germany, 1700–1815.* Ithaca, NY: Cornell University Press, 1996.

Hulme, Peter, and Neil L. Whitehead, eds. *Wild Majesty: Encounters with Caribs from Columbus to the Present Day: An Anthology*. Oxford: Clarendon Press, 1992.

Humboldt, Alexander von, and Aimé Bonpland. *Relation historique du voyage aux régions équi-noxiales du Nouveau Continent: Fait en 1799, 1800, 1801, 1802, 1803 et 1804 par Al. de Humboldt et A. Bonpland, rédigé par Alexandre de Humboldt*. Stuttgart: F. A. Brockhaus, 1970.

Humboldt, Wilhelm von. *Gesammelte Schriften*. Edited by Albert Leitzmann, Bruno Gebhardt, and Wilhelm Richter. 17 vols. Berlin: de Gruyter, 1968.

Humboldt, Wilhelm von. *Lettre à m. Abel-Rémusat, sur la nature des formes grammaticales en général, et sur le génie de la langue chinoise en particulier*. Paris: Dondey-Dupré, 1827.

Humboldt, Wilhelm von. *On Language: On the Diversity of Human Language Construction and Its Influence on the Mental Development of the Human Species*. Edited by Michael Losonsky, translated by Peter Heath. Cambridge: Cambridge University Press, 1999.

Humboldt, Wilhelm von. "On Thinking and Speaking: Sixteen Theses on Language (1795/6)." Translated by Kurt Mueller-Vollmer, 193–214. In Kurt Mueller-Vollmer, "Thinking and Speaking: Herder, Humboldt and Saussurean Semiotics." *Comparative Criticism* 11 (1989): 159–214.

Humboldt, Wilhelm von. *Prüfung der Untersuchungen über die Urbewohner Hispaniens vermittelst der vaskischen Sprache*. Berlin: Ferdinand Dümmler, 1821.

Humboldt, Wilhelm von. *Werke in fünf Bänden*. Edited by Andreas Flitner and Klaus Giel. 5 vols. Darmstadt: Wissenschaftliche Buchgesellschaft, 1960–81.

Hunt, Margaret R. *Women in Eighteenth-Century Europe*. Harlow: Pearson Longman, 2010.

Iamartino, Giovanni. "Words by Women, Words on Women in Samuel Johnson's *Dictionary of the English Language*." In *Adventuring in Dictionaries: New Studies in the History of Lexicography*, edited by John Considine, 94–125. Cambridge: Cambridge Scholars, 2010.

Inoue, Miyako. *Vicarious Language: Gender and Linguistic Modernity in Japan*. Berkeley: University of California Press, 2006.

Irigaray, Luce. *This Sex Which Is Not One*. Translated by Catherine Porter with Carolyn Burke. Ithaca, NY: Cornell University Press, 1985.

Jahn, Friedrich Ludwig, and Ernst Eiselen. *Die deutsche Turnkunst zur Einrichtung der Turnplätze*. Berlin, 1816.

Jastal, Katarzyna. "Körper und Geschlecht der deutschen Sprache im sprachnationalen Diskurs des 19. Jahrhunderts." In *Im Clash der Identitäten: Nationalismen im literatur- und kulturgeschichtlichen Diskurs*, edited by Wolfgang Brylla and Cezary Lipiński, 93–104. Göttingen: V&R Unipress, 2020.

Jespersen, Otto. "The Woman." In *Language: Its Nature, Development and Origin*, 237–54. New York: Henry Holt, 1922.

Johnson, Samuel. *Dictionary of the English Language*. 2 vols. London: J. & P. Knapton, 1755.

Jones, William Jervis. *Images of Language: Six Essays on German Attitudes to European Languages from 1500 to 1800*. Amsterdam: John Benjamins, 1999.

Jones, William Jervis. "Lingua teutonum victrix? Landmarks in German Lexicography (1500–1700)." *Histoire Epistémologie Langage* 13, no. 2 (1991): 131–52.

Juvenal. *The Satires of Juvenal*. With introduction and notes by A. F. Cole. London: J. M. Dent, 1906.

Kant, Immanuel. "Conjectural Beginning of Human History." Translated by Allen A. Wood. In *Anthropology, History, and Education*, edited by Günter Zöller and Robert B. Louden, 163–75. Cambridge: Cambridge University Press, 2007.

Kant, Immanuel. "Mutmaßlicher Anfang der Menschengeschichte." In *Abhandlungen nach 1781*, 110–23. Vol. 8 of *Gesammelte Schriften (Akademie-Ausgabe): Electronic Edition*. Charlottesville, VA: InteLex, 1999.

Kant, Immanuel. "Of the Different Human Races (1777)." In *Kant and the Concept of Race: Late Eighteenth-Century Writings*, edited by Jon M. Mikkelsen, 55–71. Albany: State University of New York Press, 2013.

Katz, Marilyn. "Ideology and the 'Status of Women' in Ancient Greece." *History and Theory* 31, no. 4 (December 1992): 70–97.

Keil, André. "'We need to rediscover our manliness . . .': The Language of Gender and Authenticity in German Right-Wing Populism." *Journal of Language and Politics* 19, no. 1 (2020): 107–24.

Key, Mary Ritchie. *Male/Female Language: With a Comprehensive Bibliography*. Metuchen, NJ: Scarecrow Press, 1975.

Kilarski, Marcin. *Nominal Classification: A History of Its Study from the Classical Period to the Present*. Amsterdam: John Benjamins, 2013.

Kim, Seongcheol. "The Populism of the Alternative for Germany (AfD): An Extended Essex School Perspective." *Palgrave Communications* 3, no. 5 (2017): 1–11.

Kittler, Friedrich. *Discourse Networks 1800/1900*. Translated by Michael Metteer with Chris Cullens. With a foreword by David E. Wellbery. Stanford, CA: Stanford University Press, 1990.

Knowles, Adam. "A Genealogy of Silence: *Chōra* and the Placelessness of Greek Women." *phi-loSOPHIA* 5, no. 1 (Winter 2015): 1–24.

Koerner, E. F. K. "Editorial: Purpose and Scope of *Historiographia Linguistica*." *Historiographia Linguistica* 1, no. 1 (1974): 1–10.

Koerner, E. F. K. "Historiography of Phonetics: The State of the Art." *Journal of the International Phonetic Association* 23, no. 1 (June 1993): 1–12.

Koerner, E. F. K. *The Importance of Techmer's Internationale Zeitschrift für Allgemeine Sprachwissenschaft in the Development of General Linguistics: An Essay*. Amsterdam: John Benjamins, 1973.

Koerner, E. F. K. "Jacob Grimm's Position in the Development of Linguistics as a Science." In *The Grimm Brothers and the Germanic Past*, edited by Elmer H. Antonsen, James Marchand, and Ladislav Zgusta, 7–23. Amsterdam: John Benjamins, 1990.

Koerner, E. F. K. "Linguistics vs. Philology: Self-Definition of a Field or Rhetorical Stance?" *Language Sciences* 19, no. 2 (1997): 167–75.

Koerner, E. F. K. "The Natural Science Impact on Theory Formation in 19th and 20th Century Linguistics." In *Professing Linguistic Historiography*, 47–76. Philadelphia: John Benjamins, 1995.

Kohler, K. "Three Trends in Phonetics: The Development of Phonetics as a Discipline in Germany since the Nineteenth Century." In *Towards a History of Phonetics*, edited by R. E. Asher and Eugénie J. A. Henderson, 161–78. Edinburgh: Edinburgh University Press, 1981.

Kolb, Alexandra. *Performing Femininity: Dance and Literature in German Modernism*. Oxford: Peter Lang, 2009.

Kolek, Vít. "Discourse of Non-Heteronormative Labelling in German-Language Press: The Case of Gendersternchen." *Slovenščina* 7, no. 2 (2019): 118–40.

Kosch, Michelle. "Agency and Self-Sufficiency in Fichte's Ethics." *Philosophy and Phenomenological Research* 91, no. 2 (September 2015): 348–80.

Kotthoff, Helga. "Gender-Sternchen, Binnen-I oder generisches Maskulinum, . . . (Akademische) Textstile der Personenreferenz als Registrierungen?" *Linguistik Online* 103, no. 3 (2020): 105–27.

Kraus, Flora. "Die Frauensprachen bei den primitiven Völkern." *Imago: Zeitschrift für Anwendung der Psychoanalyse auf die Geisteswissenschaften* 10 (1924): 296–313.

Kristeva, Julia. *Revolution in Poetic Language*. New York: Columbia University Press, 1984.

Kröll, Heinz. "Michaëlis de Vasconcellos, Carolina." *Neue Deutsche Biographie* 17 (1994): 437–38. https://www.deutsche-biographie.de/pnd119282585.html#ndbcontent.

Krug-Richter, Barbara. "'Weibergeschwätz'? Zur Geschlechtsspezifik des Geredes in der Frühen Neuzeit." In *Weibliche Rede—Rhetorik der Weiblichkeit: Studien zum Verhältnis von Rhetorik und Geschlechterdifferenz*, edited by Doerte Bischoff and Martina Wagner-Egelhaaf. Freiburg i.Br.: Rombach, 2003.

Lafitau, Joseph-François. *Moeurs des sauvages amériquains comparées aux moeurs des premiers temps*. 2 vols. Paris: Saugrain l'aîné, Charles-Estienne Hochereau, 1724.

Lakoff, Robin. "Language and Women's Place." *Language and Society* 2, no. 1 (April 1973): 45–80.

Lankes, Ana. "In Argentina, One of the World's First Bans on Gender-Neutral Language." *New York Times*, July 20, 2022.

Laqueur, Thomas Walter. *Making Sex: Body and Gender from the Greeks to Freud*. Cambridge, MA: Harvard University Press, 1992.

Lardinois, André, and Laura McClure, eds. *Making Silence Speak: Women's Voices in Greek Literature and Society*. Princeton, NJ: Princeton University Press, 2001.

Lasch, Richard. "Über Sondersprachen und ihre Entstehung." *Mitteilungen der Anthropologischen Gesellschaft in Wien* 37 (1907): 89–101.

Lawson, Robert. *Language and Mediated Masculinities: Cultures, Contexts, Constraints*. Oxford: Oxford University Press, 2023.

Le Doeuff, Michèle. *Hipparchia's Choice: An Essay Concerning Women, Philosophy, Etc.* Translated by Trista Selous. Oxford: Blackwell, 1991.

Lefmann, Salomon. *August Schleicher: Skizze*. Leipzig: B. G. Teubner, 1870.

Lenik, Stephan. "Carib as a Colonial Category: Comparing Ethnohistoric and Archaeological Evidence from Dominica, West Indies." *Ethnohistory* 59, no. 1 (Winter 2012): 79–107.

Lepschy, Giulio C., ed. *History of Linguistics*. 4 vols. London: Pearson Education, 1994.

Lepsius, Karl Richard. *Nubische Grammatik*. Berlin: Wilhelm Hertz, 1880.

Lepsius, Karl Richard. *Standard Alphabet for Reducing Unwritten Languages and Foreign Graphic Systems to a Uniform Orthography in European Letters*. 2nd ed. London: Williams & Norgate, 1863.

Lifschitz, Avi. *Language and Enlightenment: The Berlin Debates of the Eighteenth Century*. Oxford: Oxford University Press, 2016.

Locke, John. *An Essay Concerning Human Understanding*. Edited by Peter H. Nidditch. Oxford: Clarendon Press, 1990.

Locke, John. *First Treatise*. In *Two Treatises of Government* and *A Letter Concerning Toleration*, edited by Ian Shapiro, with essays by John Dunn, Ruth W. Grant, and Ian Shapiro, 7–99. New Haven, CT: Yale University Press, 2003.

Lubrich, Oliver. "Alexander von Humboldt: Revolutionizing Travel Literature." *Monatshefte* 96, no. 3 (Fall 2004): 360–87.

Lucian. *Lucian, Volume 5*. Translated by A. M. Harmon. Loeb Classical Library 302. Cambridge, MA: Harvard University Press, 1936.

Lucian. *Lucian, Works.* Vol. 7. Translated by M. D. MacLeod. Loeb Classical Library 437. Cambridge, MA: Harvard University Press, 1961.

Maaler, Josua. *Die Teütsch spraach. Alle wuerter/ namen/ uñ arten zü reden in Hochteütscher spraach.* Tiguri: Christoph Froschoverus, 1561.

Maas, Utz. "Die erste Generation der deutschsprachigen Sprachwissenschaftlerinnen." *STUF— Language Typology and Universals* 44, no. 1 (December 1991): 61–69.

MacLeod, Catriona. *Embodying Ambiguity: Androgyny and Aesthetics from Winckelmann to Keller.* Detroit: Wayne State University Press, 1998.

Mander, Jenny. "Marivaux's *Paysan parvenu* and the Genre of Conjectural History." *Nottingham French Studies* 48, no. 3 (Autumn 2009): 20–30.

Mandeville, Bernard de. *The Fable of the Bees: or, Private Vices, Publick Benefits.* Edited by F. B. Kaye. 2 vols. Oxford: Oxford University Press, 2014.

Martens, Lorna. "The Theme of the Repressed Memory in Hofmannsthal's *Elektra.*" *German Quarterly* 60, no. 1 (Winter 1987): 38–51.

Masten, Jeffrey. *Queer Philologies: Sex, Language, and Affect in Shakespeare's Time.* Philadelphia: University of Pennsylvania Press, 2016.

Mauthner, Fritz. *Beiträge zu einer Kritik der Sprache.* 3rd ed. 3 vols. Leipzig, 1923. Reprint, Hildesheim: Georg Olms, 1967–69.

Mazón, Patricia M. *Gender and the Modern Research University: The Admission of Women to German Higher Education, 1865–1914.* Stanford, CA: Stanford University Press, 2003.

McBride, Patrizia. *The Void of Ethics: Robert Musil and the Experience of Modernity.* Evanston, IL: Northwestern University Press, 2006.

McConnell-Ginet, Sally. "Gender and Its Relation to Sex: The Myth of 'Natural' Gender." In *The Expression of Gender,* edited by Greville G. Corbett, 3–38. Berlin: de Gruyter Mouton, 2014.

McConnell-Ginet, Sally. "Intonation in a Man's World." *Signs* 3, no. 3 (Spring 1978): 541–59.

McConnell-Ginet, Sally. "Language and Gender." In *Language: The Socio-Cultural Context,* 75–99. Vol. 4 of *Linguistics: The Cambridge Survey,* edited by Frederick J. Newmeyer. Cambridge: Cambridge University Press, 1988.

Meiner, Johann Werner. *Versuch einer an der menschlichen Sprache abgebildeten Vernunftlehre: oder Philosophische und allgemeine Sprachlehre.* Leipzig: Breitkopf, 1781.

Meiners, Christoph. *Geschichte des weiblichen Geschlechts.* 4 vols. Hannover: Helwingsche Hofbuchhandlung, 1788.

Menand, Louis, Paul Reitter, and Chad Wellmon, eds. *The Rise of the Research University: A Sourcebook.* Chicago: University of Chicago Press, 2017.

Mendicino, Kristina. *Prophecies of Language: The Confusion of Tongues in German Romanticism.* New York: Fordham University Press, 2016.

Messling, Markus. "Text and Determination: On Racism in 19th Century European Philology." *Philological Encounters* 1 (2016): 79–104.

Meyer Spacks, Patricia. *The Female Imagination.* New York: Knopf, 1975.

Michaelis, Johann David. *A Dissertation on the Influence of Opinions on Language and of Language on Opinions, which Gained the Prussian Royal Academy's Prize on that Subject.* London: W. Owen and W. Bingley, 1769.

Michaelis, Johann David. *Dissertation qui a remporté le prix proposé par l'Academie royale des sciences et belles lettres de Prusse, sur l'influence réciproque du langage sur les opinions, et des opinions sur le langages.* Berlin: Haude & Spener, 1760.

Michaelis, Johann David. "Preisschrift vom Ursprung der Sprache, so ich am 12. Dec. 1770 an die Akademie zu Berlin gesandt habe." Unpublished manuscript. Codex Michaelis 72. Niedersächsisches Staats- und Universitätsbibliothek Göttingen Abteilung Handschriften und Alte Drucke Nachlass Michaelis.

Michaelson, Patricia Howell. *Speaking Volumes: Women, Reading, and Speech in the Age of Austen*. Stanford, CA: Stanford University Press, 2002.

Milton, John. *Paradise Lost*. Edited by Gordon Teskey. New York: W. W. Norton, 2005.

Moffitt, John F., and Santiago Sebastián. *O Brave New People: The European Invention of the American Indian*. Albuquerque: University of New Mexico Press, 1998.

Moland, Lydia L. "Conjectural Truths: Kant and Schiller on Educating Humanity." In *Aesthetics, History, Politics, and Religion*, 91–108. Vol. 2 of *Kant and His German Contemporaries*, edited by Daniel O. Dahlstrom. Cambridge: Cambridge University Press, 2018.

Monboddo, Lord James Burnett. *Of the Origin and Progress of Language*. 2nd ed. 2 vols. Edinburgh: J. Balfour, 1774.

Moore, Francis. *Travels Into the Inland Parts of Africa: Containing a Description of the Several Nations for the Space of Six Hundred Miles Up the River Gambia* [. . .]. London: Edward Cave, 1738.

Moritz, Karl Philipp. *Deutsche Sprachlehre für die Damen: In Briefen*. Berlin: Arnold Wever, 1782.

Morpurgo Davies, Anna. *Nineteenth-Century Linguistics*. Vol. 4 of *History of Linguistics*, edited by Giulio C. Lepschy. London: Routledge, 1998.

Morrison, Toni. *Playing in the Dark: Whiteness and the Literary Imagination*. Cambridge, MA: Harvard University Press, 1992.

Muchembled, Robert. *Smells: A Cultural History of Odours in Early Modern Times*. Translated by Susan Pickford. Cambridge: Polity Press, 2020.

Müller, Margarethe. *Carla Wenckebach, Pioneer*. Boston: Ginn, 1908.

Müller, Max. *Chips from a German Workshop*. 4 vols. London: Longmans, Green, & Co., 1867.

Müller-Sievers, Helmut. *The Science of Literature: Essays on an Incalculable Difference*. Translated by Chadwick Truscott, Paul Babinski, and Helmut Müller-Sievers. Berlin: de Gruyter, 2015.

Müller-Sievers, Helmut. *Self-Generation: Biology, Philosophy, and Literature around 1800*. Stanford, CA: Stanford University Press, 1997.

Münchhausen, Börries von. *Clementine von Münchhausen, geb. v. d. Gabelentz: Anlagen u. Talente, Entwicklung u. Studien*. Unpublished manuscript, 1907. Deutsche Nationalbibliothek Leipzig. Typescript.

Münchhausen, Clementine von. "H. Georg v. d. Gabelentz. Biographie und Charakteristik [1913]," edited by Annemete von Vogel. In *Georg von der Gabelentz: Ein biographisches Lesebuch*, edited by Kennosuke Ezawa and Annemete von Vogel, 85–171. Tübingen: Gunter Narr, 2013.

Murner, Thomas. "Das klapper benckly." In *Schriften*, edited by Franz Schultz, vol. 3, 88–89. Berlin: de Gruyter, 1925.

Musil, Robert. *Five Women*. Translated by Eithne Wilkins and Ernst Kaiser. Boston: Verba Mundi, 1999.

Musil, Robert. *Gesammelte Werke*. Edited by Adolf Frisé. 9 vols. Reinbek bei Hamburg: Rowohlt, 1978.

Musil, Robert. *Intimate Ties: Two Novellas*. Translated by Peter Wortsman. New York: Archipelago Books, 2019.

Musil, Robert. *Precision and Soul: Essays and Addresses*. Edited and translated by Burton Pike and David S. Luft. Chicago: University of Chicago Press, 1990.

Musil, Robert. *Tagebücher, Aphorismen, Essays und Reden*. Hamburg: Rowohlt, 1955.

Musil, Robert. *Young Törless*. Translated by Eithne Wilkins and Ernst Kaiser. London: Secker & Warburg, 1955.

Nadelkunst der Clementine von Münchhausen, geb. von der Gabelentz, 1849–1913. Wunstorf: A. Jacques, 2000.

Neis, Cordula. *Anthropologie im Sprachdenken des 18. Jahrhunderts: Die Berliner Preisfrage nach dem Ursprung der Sprache*. Berlin: de Gruyter, 2003.

Neuman, Nichole M. "Illi Anna Heger's Grammatical Futurity." *Seminar* 56, no. 3–4 (November 2020): 302–21.

The New Oxford Annotated Bible: New Revised Standard Edition. Edited by Michael D. Coogan. Oxford: Oxford University Press, 2010.

Nicot, Jean. *Le Thresor de la langue francoyse, tant ancienne que moderne*. Paris: David Douceur, 1606.

Nietzsche, Friedrich. "On Truth and Lies in a Nonmoral Sense" [1873]. In *The Nietzsche Reader*, edited by Keith Ansell Pearson and Duncan Large, 114–23. Malden, MA: Blackwell, 2006.

Norberg, Jakob. *The Brothers Grimm and the Making of German Nationalism*. Cambridge: Cambridge University Press, 2022.

North, Helen F. "The Mare, the Vixen, and the Bee: 'Sophrosyne' as the Virtue of Women in Antiquity." *Illinois Classical Studies* 2 (1977): 35–48.

North, Helen F. *Sophrosyne: Self-Knowledge and Self-Restraint in Greek Literature*. Ithaca, NY: Cornell University Press, 1966.

Nübel, Birgit. "Krähende Hühner und gelehrte Weiber: Aspekte des Frauenbildes bei Johann Gottfried Herder." In *Herder Jahrbuch*, edited by Wulf Koepke and Wilfried Malsch, 29–49. Stuttgart: J. B. Metzler, 1994.

Nye, Robert A. "How Sex Became Gender." *Psychoanalysis and History* 12, no. 2 (2010): 195–209.

Nyland, Chris. "Adam Smith, Stage Theory, and the Status of Women." In *The Status of Women in Classical Economic Thought*, edited by Robert Dimand and Chris Nyland, 86–107. Northampton, MA: Edward Elgar, 2003.

O'Brien, Karen. "From Savage to Scotswoman: The History of Femininity." In *Women and the Enlightenment in Eighteenth-Century Britain*, 68–109. Cambridge: Cambridge University Press, 2009.

Offen, Karen. *European Feminisms 1700–1950: A Political History*. Stanford, CA: Stanford University Press, 2000.

Oldstone-Moore, Christopher. *Of Beards and Men: The Revealing History of Facial Hair*. Chicago: University of Chicago Press, 2015.

Olender, Maurice. *The Languages of Paradise: Race, Religion, and Philology in the Nineteenth Century*. Translated by Arthur Goldhammer. Cambridge, MA: Harvard University Press, 1992.

Olivier, Jacques [Alexis Trousset]. *Alphabet de l'imperfection et malice des femmes*. Paris: Jean Petit-Pas, 1617.

Oliver, Kelly, and Lisa Walsh, eds. *Contemporary French Feminism*. Oxford: Oxford University Press, 2004.

Osthoff, Hermann, and Karl Brugmann. *Morphologische Untersuchungen auf dem Gebiete der indogermanischen Sprachen*. Leipzig: S. Hirzel, 1878.

Oswald, J. H. *Das grammatische Geschlecht und seine sprachliche Bedeutung.* Paderborn: Junfer-mann, 1866.

Ovid. *Metamorphoses: A New Verse Translation.* Translated by David Raeburn. London: Pen-guin, 2004.

Palmeri, Frank. *Stages of Nature, Stages of Society: Enlightenment Conjectural History and Mod-ern Social Discourse.* New York: Columbia University Press, 2016.

Parker, Patricia. *Literary Fat Ladies: Rhetoric, Gender, Property.* London: Methuen, 1987.

Paul, Hermann. *Principien der Sprachgeschichte.* Halle: Max Niemeyer, 1886.

Paul, Hermann. *Principles of the History of Language.* Translated by H. A. Strong. London: Swan Sonnenschein, Lowrey & Co., 1888.

Pauwels, Jacques R. *Women, Nazis, and Universities: Female University Students in the Third Reich, 1933–1945.* Westport, CT: Greenwood Press, 1984.

Peaden, Catherine Hobbs. "Condillac and the History of Rhetoric." *Rhetorica: A Journal of the History of Rhetoric* 11, no. 2 (Spring 1993): 135–56.

Petrella, Sara. "Femmes à poils: Réception et actualisation d'un cliché dans les *Moeurs des sau-vages ameriquains* de Joseph-François Lafitau." In *La Plume et le calumet: Joseph-François Lafitau et les "sauvages ameriquains,"* edited by Mélanie Lozat and Sara Petrella, 139–51. Paris: Garnier, 2019.

Pettigrew, William A. *Freedom's Debt: The Royal African Company and the Politics of the Atlantic Slave Trade, 1672–1752.* Chapel Hill: Omohundro Institute and University of North Carolina Press, 2013.

Piser, Karina. "Aux Armes, Citoyen·nes!" *Foreign Policy,* July 4, 2021.

Plato. *Cratylus.* Translated by C. D. C. Reeve. In *Complete Works,* edited by John M. Cooper. Indianapolis: Hackett, 1997.

Ploss, Heinrich, and Max Bartels. *Das Weib in der Natur- und Völkerkunde.* Edited by Paul Bar-tels. 2 vols. Leipzig: Th. Grieben, 1908.

Pollock, Sheldon, Benjamin A. Elman, and Ku-ming Kevin Chang, eds. *World Philology.* Cam-bridge, MA: Harvard University Press, 2015.

Pott, August. "Metaphern, vom leben und von körperlichen lebensverrichtungen hergenommen." *Zeitschrift für vergleichende Sprachforschung auf dem Gebiete des Deutschen, Griechischen und Lateinischen* 2 (1853): 101–27.

Pourciau, Sarah. *The Writing of Spirit: Soul, System, and the Roots of Language Science.* New York: Fordham University Press, 2017.

Pusch, Luise F. *Alle Menschen werden Schwestern.* Frankfurt a.M.: Suhrkamp, 1990.

Raggam-Blesch, Michaela. "A Pioneer in Academia: Elise Richter." In *Jewish Intellectual Women in Central Europe, 1860–2000: Twelve Biographical Essays,* edited by Judith Szapor, Andrea Pető, and Marina Calloni, 93–128. Lewiston, NY: Edwin Mellen, 2012.

Rancière, Jacques. *The Names of History: On the Poetics of Knowledge.* Translated by Hassan Melehy. Minneapolis: University of Minnesota Press, 1994.

Rausch, Sven. "Sophrosyne." In *Der Neue Pauly: Enzyklopädie der Antike,* edited by Hubert Can-cik, Helmuth Schneider, and Manfred Landfester. First published online 2006. http://dx.doi .org/10.1163/1574-9347_bnp_e1117400.

Reill, Peter Hanns. "Science and the Construction of the Cultural Sciences in Late Enlighten-ment Germany: The Case of Wilhelm von Humboldt." *History and Theory* 33, no. 3 (Octo-ber 1994): 345–66.

Reill, Peter Hanns. "The Scientific Construction of Gender and Generation in the German Late

Enlightenment and in German Romantic *Naturphilosophie*." In *Reproduction, Race, and Gender in Philosophy and the Early Life Sciences*, edited by Susanne Lettow, 65–82. Albany: State University of New York Press, 2014.

Richardson, Sarah S. *Sex Itself: The Search for Male and Female in the Human Genome*. Chicago: University of Chicago Press, 2013.

Richter, Elise. "Erziehung und Entwicklung." In *Führende Frauen Europas in sechzehn Selbstschilderungen*, edited by Elga Kern, 70–93. Munich: E. Reinhardt, 1928.

Richter, Elise. "Erziehung und Entwicklung." In *Kleinere Schriften zur allgemeinen und romanischen Sprachwissenschaft*, edited by Yakov Malkiel, 521–54. Innsbruck: H. Kowatsch, 1977.

Richter, Elise. *Wie wir sprechen: Sechs volkstümliche Vorträge*. Leipzig: B. G. Teubner, 1912.

Richter, Simon. "Weimar Heteroclassicism: Wilhelm von Humboldt, Caroline von Wolzogen, and the Aesthetics of Gender." *Publications of the English Goethe Society* 81, no. 3 (October 2012): 137–51.

Rinne, Johann Karl Friedrich. *Die natürliche Entstehung der Sprache aus dem Gesichtspuncte der historischen oder vergleichenden Sprachwissenschaft*. Erfurt: Friedrick Wilhelm Otto, 1834.

Robertson, Ritchie. "'Ich habe ihm das Beil nicht geben können': The Heroine's Failure in Hofmannsthal's *Elektra*." *Orbis Litterarum* 41 (1986): 312–31.

Robinson, Orrin W. *Grimm Language: Grammar, Gender and Genuineness in the Fairy Tales*. Philadelphia: John Benjamins, 2010.

Rochefort, Charles de. *Histoire naturelle et morale des îles Antilles de l'Amérique: Enrichie de plusieurs belles figures des raretez les plus considerables qui y sont d'écrites, avec un vocabulaire Caraïbe*. Rotterdam: Arnould Leers, 1658.

Rochefort, Charles de. *Historische Beschreibung der Antillen Inseln in America gelegen: In sich begreiffend deroselben Gelegenheit, darinnen befindlichen natürlichen Sachen, sampt deren Einwohner Sitten und Gebräuchen*. 2 vols. Frankfurt a.M.: Wilhelm Serlins, 1668.

Roethe, Gustav. "Zum neuen Abdruck." In Jacob Grimm, *Deutsche Grammatik*, edited by Gustav Roethe and Edward Schröder, vol. 3, ix–xxxi. Gütersloh: C. Bertelsmann, 1890.

Roman, Luke, and Monica Roman, eds. *Encyclopedia of Greek and Roman Mythology*. New York: Facts on File, 2010.

Römer, Ruth. *Sprachwissenschaft und Rassenideologie in Deutschland*. Munich: Wilhelm Fink, 1985.

Rousseau, Jean-Jacques. *Discourse on Inequality: On the Origin and Basis of Inequality among Men*. Auckland: Floating Press, 2009.

Rousseau, Jean-Jacques. *Discourse on the Origin of Inequality*. Edited by Patrick Coleman. Translated by Franklin Philip. Oxford: Oxford University Press, 1994.

Rousseau, Jean-Jacques. *Discours sur l'origine et les fondements de l'inégalité parmi les hommes*. Paris: Gallimard, 2006.

Rousseau, Jean-Jacques. *Emile, or On Education*. Translated by Allan Bloom. New York: Basic Books, 1979.

Rousseau, Jean-Jacques. *Émile, ou de l'éducation*. Paris: Garnier-Flammarion, 1966.

Rousseau, Jean-Jacques. *Essai sur l'origine des langues, où il est parlé de la mélodie et de l'imitation musicale*. Edited by Charles Porset. Bordeaux: Ducros, 1970

Rousseau, Jean-Jacques. "Essay on the Origin of Language." In Jean-Jacques Rousseau and Johann Gottfried Herder, *On the Origin of Language*, translated by John H. Moran and Alexander Gode, 5–84. Chicago: University of Chicago Press, 1986.

Roux, Benoît. "Le pasteur Charles de Rochefort et l'*Histoire naturelle et morale des îles Antilles de l'Amérique*." *Cahiers d'Histoire de l'Amérique Coloniale* 5 (2011): 175–216.

Royen, Gerlach. *Die nominalen Klassifikations-Systeme in den Sprachen der Erde.* Mödling: Anthropos, 1929.

Russell, Lindsay Rose. *Women and Dictionary-Making: Gender, Genre, and English Language Lexicography.* Cambridge: Cambridge University Press, 2018.

Salvo, Sophie. "Father Is Always Uncertain: J. J. Bachofen and the Epistemology of Patriarchy." *Monatshefte* 116, no. 1 (Winter 2024): 44–65.

Saussure, Ferdinand de. *Course in General Linguistics.* Edited by Perry Meisel and Haun Saussy. Translated by Wade Baskin. New York: Columbia University Press, 2011.

Schiebinger, Londa. *The Mind Has No Sex? Women in the Origins of Modern Science.* Cambridge, MA: Harvard University Press, 1991.

Schiebinger, Londa. *Nature's Body: Gender in the Making of Modern Science.* New Brunswick, NJ: Rutgers University Press, 2004.

Schiller, Friedrich. "Über die ästhetische Erziehung des Menschen in einer Reihe von Briefen." In *Philosophische Schriften I*, edited by Benno von Wiese, 309–412. Vol. 20 of *Schillers Werke: Nationalausgabe.* Weimar: Hermann Böhlaus Nachfolger, 1962.

Schlapbach, Karin. "Lucian's *On Dancing* and the Models for a Discourse on Pantomime." In *New Directions in Ancient Pantomime*, edited by Edith Hall and Rose Wyles, 314–37. Oxford: Oxford University Press, 2008.

Schlegel, Friedrich. *On the Language and Philosophy of the Indians.* In *The Aesthetic and Miscellaneous Works*, translated by E. J. Millington, 424–626. London: Henry G. Bohn, 1849.

Schlegel, Friedrich. *Über die Sprache und Weisheit der Indier* [1808]. In *Studien zur Philosophie und Theologie 1796–1824*, edited by Ernst Behler and Ursula Struc-Oppenberg, 105–433. Vol. 8 of *Kritische Friedrich-Schlegel-Ausgabe.* Munich: Ferdinand Schöningh, 1975.

Schleicher, August. *Die deutsche Sprache.* Stuttgart: Cotta, 1860.

Schleicher, August. "Linguistik und Philologie." In *Die Sprachen Europas in systematischer Uebersicht*, 1–5. Bonn: H. B. König, 1850.

Schmeling, Manfred, and Kerst Walstra. "Erzählung." In *A–G*, edited by Klaus Weimar, 517–22. Vol. 1 of *Reallexikon der deutschen Literaturwissenschaft*, edited by Georg Braungart, Harald Fricke, Klaus Grubmüller, Jan-Dirk Müller, Friedrich Vollhardt, and Klaus Weimar. 3rd ed. Berlin: de Gruyter, 2007.

Schnicke, Falko. *Die männliche Disziplin: Zur Vergeschlechtlichung der deutschen Geschichtswissenschaft 1780–1900.* Göttingen: Wallstein, 2015.

Schottelius, J. G. *Teutsche Sprachkunst.* Braunschweig: Gruber, 1641.

Schröder, Jürgen. "Am Grenzwert der Sprache: Zu Robert Musils 'Vereinigungen.'" *Euphorion* 60 (1966): 311–34.

Schwarz, F. H. C. *Grundsätze der Töchtererziehung für die Gebildeten.* Jena: Cröker, 1836.

Scott, Jill. "From Pathology to Performance: Hugo von Hofmannsthal's *Elektra* and Sigmund Freud's 'Fräulein Anna O.'" In *Electra after Freud: Myth and Culture*, 57–80. Ithaca, NY: Cornell University Press, 2005.

Scott, Joan Wallach. *Only Paradoxes to Offer: French Feminists and the Rights of Man.* Cambridge, MA: Harvard University Press, 1996.

Seifert, Nicole. *Frauen Literatur: Abgewertet, vergessen, wiederentdeckt.* Cologne: Kiepenheuer & Witsch, 2021.

Semonides. "Fragments." In *Greek Iambic Poetry: From the Seventh to the Fifth Centuries BC*, edited and translated by Douglas E. Gerber, 298–341. Loeb Classical Library 259. Cambridge, MA: Harvard University Press, 1999.

Showalter, Elaine. "Towards a Feminist Poetics." In *Women Writing and Writing about Women*, edited by Mary Jacobus, 22–41. London: Croom Helm, 1979.

Simrock, Karl. *Die Deutschen Sprichwörter gesammelt*. Frankfurt a.M.: H. L. Brönner, 1846.

Skotnes, Pippa, ed. *Claim to the Country: The Archive of Lucy Lloyd and Wilhelm Bleek*. Johannesburg: Jacana, 2007.

Smith, Adam. "Considerations Concerning the Formation of Languages." In *The Works of Adam Smith*, edited by Dugald Stewart, vol. 5, 1–48. London: T. Cadell, 1811.

Smith, Bonnie G. *The Gender of History: Men, Women, and Historical Practice*. Cambridge, MA: Harvard University Press, 1998.

Smith-Prei, Carrie, and Maria Stehle. *Awkward Politics: Technologies of Popfeminist Activism*. Montreal: McGill–Queen's University Press, 2016.

Solanki, Tanvi. "Aural Philology: Herder Hears Homer Singing." *Classical Receptions Journal* 12, no. 4 (October 2020): 401–24.

Spender, Dale. *Man Made Language*. London: Routledge, 1980.

Stam, James H. *Inquiries into the Origin of Language: The Fate of a Question*. New York: Harper & Row, 1976.

Stavreva, Kirilka. *Words Like Daggers: Violent Female Speech in Early Modern England*. Lincoln: University of Nebraska Press, 2015.

Steinbrügge, Lieselotte. *The Moral Sex: Woman's Nature in the French Enlightenment*. Translated by Pamela E. Selwyn. Oxford: Oxford University Press, 1995.

Steinthal, Heymann. *Abriss der Sprachwissenschaft: Einleitung in die Psychologie und Sprachwissenschaft*. Berlin: Ferd. Dümmler, 1871.

Steinthal, Heymann. *Geschichte der Sprachwissenschaft bei den Griechen und Römern mit besonderer Rücksicht auf die Logik*. Berlin: Ferd. Dümmler, 1863.

Stepan, Nancy Leys. "Race and Gender: The Role of Analogy in Science." *Isis* 77, no. 2 (June 1986): 261–77.

Stewart, Dugald. "Account of the Life and Writings of Adam Smith, LL. D." In *The Collected Works of Dugald Stewart*, edited by William Hamilton, vol. 10, 1–100. Edinburgh: Thomas Constable, 1858.

Stieler, Kaspar von. *Der teutschen Sprache Stammbaum und Fortwachs, oder Teutscher Sprachschatz*. Nürnberg: Johann Hofmann, 1691.

Stoll, Otto. *Zur Ethnographie der Republik Guatemala*. Zürich: Orell Füssli, 1884.

Strube, Sonja A., Rita Perintfalvi, Raphaela Hemet, Miriam Metze, and Cicek Sahbaz, eds. *Anti-Genderismus in Europa: Allianzen von Rechtspopulismus und religiösem Fundamentalismus*. Bielefeld: Transcript, 2021.

Surber, Jere Paul. *Language and German Idealism: Fichte's Linguistic Philosophy*. Atlantic Highlands, NJ: Humanities Press, 1996.

Süssmilch, Johann Peter. *Versuch eines Beweises, dass die erste Sprache ihren Ursprung nicht von Menschen, sondern allein vom Schöpfer erhalten habe*. Berlin: Buchladen der Realschule, 1766.

Sweet, Paul Robinson. *Wilhelm von Humboldt: A Biography*. 2 vols. Columbus: Ohio State University Press, 1980.

Swiggers, Pierre. "A Note on the History of the Term *Linguistics*. With a Letter from Peter Stephen Du Ponceau to Joseph von Hammer-Purgstall." *Beiträge zur Geschichte der Sprachwissenschaft* 6 (1996): 1–17.

Sylvain, Cabela de, and Carsten Balzer. "Die SYLVAIN-Konventionen—Versuch einer 'geschlechtergerechten' Grammatik-Transformation der deutschen Sprache." *Liminalis* 2 (2008): 40–53.

Tannen, Deborah. *Talking from 9 to 5: Women and Men at Work*. New York: Avon, 1994.

Tannen, Deborah. *You Just Don't Understand: Women and Men in Conversation*. New York: Ballantine Books, 1990.

Taylor, Douglas R. "Languages and Ghost-Languages of the West-Indies." *International Journal of American Linguistics* 22, no. 2 (April 1956): 180–83.

Taylor, Douglas R., and Berend J. Hoff. "The Linguistic Repertory of the Island-Carib in the Seventeenth Century: The Men's Language: A Carib Pidgin?" *International Journal of American Linguistics* 46, no. 4 (October 1980): 301–12.

Techmer, Friedrich. *Phonetik: Zur vergleichenden Physiologie der Stimme und Sprache*. 2 vols. Leipzig: Wilhelm Engelmann, 1880.

Timm, Annette F., and Joshua A. Sanborn. *Gender, Sex and the Shaping of Modern Europe: A History from the French Revolution to the Present Day*. 3rd ed. London: Bloomsbury, 2022.

Toews, John Edward. *Becoming Historical: Cultural Reformation and Public Memory in Early Nineteenth-Century Berlin*. Cambridge: Cambridge University Press, 2004.

Tonger-Erk, Lily. *Actio: Körper und Geschlecht in der Rhetoriklehre*. Berlin: de Gruyter, 2012.

Townson, Michael. *Mother-Tongue and Fatherland: Language and Politics in German*. Manchester: Manchester University Press, 1992.

Trabant, Jürgen. *Apeliotes, oder, Der Sinn der Sprache: Wilhelm von Humboldts Sprach-Bild*. Munich: W. Fink, 1986.

Trabant, Jürgen. "Es knackt im Gebälk der Republik." *Die Welt*, May 17, 2021.

Trabant, Jürgen. "Introduction: New Perspectives on an Old Academic Question." In *New Essays on the Origin of Language*, edited by Jürgen Trabant and Sean Ward, 1–17. New York: Mouton de Gruyter, 2001.

Trabant, Jürgen. "Nachwort." In Wilhelm von Humboldt, *Über die Sprache*, 201–17. Tübingen: A. Francke, 1994.

Trömel-Plötz, Senta. *Frauensprache: Sprache der Veränderung*. Frankfurt a.M.: Fischer, 1982.

Tyson, Sarah. *Where are the Women? Why Expanding the Archive Makes Philosophy Better*. New York: Columbia University Press, 2018.

Underhill, James W. *Humboldt, Worldview and Language*. Edinburgh: Edinburgh University Press, 2009.

Vasconcellos, Carolina Michaëlis de. *Studien zur romanischen Wortschöpfung*. Leipzig: F. A. Brockhaus, 1876.

Veit-Brause, Irmline. "Scientists and the Cultural Politics of Academic Disciplines in Late 19th-Century Germany: Emil Du Bois-Reymond and the Controversy over the Role of the Cultural Sciences." *History of the Human Sciences* 14, no. 4 (November 2001): 31–56.

Veldre, Georgia. "Zur Diskussion über den Begriff 'Tochtersprache' im 19. Jahrhundert." *Historiographia Linguistica* 19, no. 1 (1992): 65–96.

Vico, Giambattista. *The New Science of Giambattista Vico*. Translated by Thomas Goddard Bergin and Max Harold Fisch. Ithaca, NY: Cornell University Press, 1984.

Vogel, Annemete von, and James McElvenny. "The Gabelentz Family in their Own Words." In *Georg von der Gabelentz and the Science of Language*, edited by James McElvenny, 13–26. Amsterdam: Amsterdam University Press, 2019.

Vogl, Joseph. "Für eine Poetologie des Wissens." In *Die Literatur und die Wissenschaften, 1770–1930*, edited by Karl Richter, Jörg Schönert, and Michael Titzmann, 107–27. Stuttgart: M & P, 1997.

Vogt, Carl. *Vorlesungen über den Menschen, seine Stellung in der Schöpfung und in der Geschichte der Erde*. Gießen: J. Ricker, 1863.

von Mücke, Dorothea. *Virtue and the Veil of Illusion: Generic Innovation and the Pedagogical Project in Eighteenth-Century Literature*. Stanford, CA: Stanford University Press, 1991.

Wagner, Richard. "Das Judenthum in der Musik." In Frank Piontek, *Richard Wagners "Das Judenthum in der Musik": Text, Kommentar und Wirkungsgeschichte*, 20–52. Beucha: Sax, 2017.

Wagner, Richard. "Judaism in Music." In *Judaism in Music and Other Essays*, translated by William Ashton Ellis, 75–122. Lincoln: University of Nebraska Press, 1995.

Waldow, Anik, and Nigel DeSouza, eds. *Herder: Philosophy and Anthropology*. Oxford: Oxford University Press, 2017.

Walther, Philipp Franz von. *Physiologie des Menschen mit durchgängiger Rücksicht auf die comparative Physiologie der Thiere*. 2 vols. Landshut: Krüll, 1807-1808.

Warburton, William. *The Divine Legation of Moses Demonstrated*. 2 vols. London: Fletcher Gyles, 1741.

Warren, Paul. *Uptalk: The Phenomenon of Rising Intonation*. Cambridge: Cambridge University Press, 2016.

Weideger, Paula. *History's Mistress: A New Interpretation of a Nineteenth-Century Ethnographic Classic*. New York: Viking Penguin, 1985.

Weigel, Sigrid. *Body- and Image-Space: Re-Reading Walter Benjamin*. Translated by Georgina Paul with Rachel McNicholl and Jeremy Gaines. London: Routledge, 1996.

Weiss, Penny A. *Gendered Community: Rousseau, Sex, and Politics*. New York: New York University Press, 1993.

Welcker, Hermann. *Untersuchungen über Wachstum und Bau des menschlichen Schädels*. Leipzig: Wilhelm Engelmann, 1862.

Wellbery, David E. "Die Opfer-Vorstellung als Quelle der Faszination: Anmerkungen zum Chandos-Brief und zur frühen Poetik Hofmannsthals." In *Hofmannsthal Jahrbuch* 11, edited by Gerhard Neumann, Ursula Renner, Günter Schnitzler, and Gotthart Wunberg, 281–310. Freiburg i.Br.: Rombach, 2003.

Wenckebach, Carla, and Josepha Schrakamp. *Deutsche Grammatik für Amerikaner: Nach einer praktischen Methode*. 4th ed. New York: F. W. Christern, 1884.

Wenckebach, Carla, and Helene Wenckebach. *Deutscher Anschauungs-Unterricht für Amerikaner: Ein Hilfsbuch*. Boston: Carl Schoenhof, 1886.

Wenckebach, Carla, and Helene Wenckebach. *Deutsches Lesebuch*. Boston: Carl Schoenhof, 1887.

Whitehead, Neil L., ed. *Wolves from the Sea: Readings in the Anthropology of the Native Caribbean*. Leiden: KITLV Press, 1995.

Wilke, Tobias. *Sound Writing: Experimental Modernism and the Poetics of Articulation*. Chicago: University of Chicago Press, 2022.

Wittgenstein, Ludwig. *Tractatus Logico-Philosophicus*. Edited by Luciano Bazzocchi. London: Anthem Press, 2022.

Wolff, Tristram. *Against the Uprooted Word: Giving Language Time in Transatlantic Romanticism*. Stanford, CA: Stanford University Press, 2022.

Wolke, Christian Heinrich. *Anleit zur deutschen Gesamtsprache oder zur Erkennung und Berichtung einiger (zu wenigst 20) tausend Sprachfehler in der hochdeutschen Mundart*. Dresden, 1812.

Woolf, Virginia. *A Room of One's Own*. In *A Room of One's Own* and *Three Guineas*, edited by Morag Shiach, 1–149. Oxford: Oxford University Press, 1992.

Zelle, Carsten. "Konstellationen der Moderne. Verstummen—Medienwechsel—literarische Phänomenologie." *Musil-Forum* 27 (2001/2002): 88–102.

Zipes, Jack. *Fairy Tales and the Art of Subversion: The Classical Genre for Children and the Process of Civilization*. New York: Routledge, 1991.

Index

Page numbers in italics indicate figures. "Grimm" throughout the index refers to Jacob Grimm, and "Humboldt" refers to Wilhelm von Humboldt.

—negative reactions to, 83, 141, 205n144 (chap. 2)
—on *Weibersprache*, 63, 80; Carib, 83–84; as deviant, 2, 57–58; Native American, 84;

imitation: distinguishes animals from humans (Herder), 31; and language acquisition (Herder), 27; pantomime as (Hofmannsthal), 144; women and, 33, 34
Indo-European languages, as "Aryan" and most advanced, 110
Irigaray, Luce, 10
Islam, 171, 174
"Island Caribs"
—as brutal savages, 64–66, 68–69
—language of, 68
—language/ethnography of, 197nn7–9 (chap. 2); women's, 92–93, 206n178 (chap. 2)
See also Arawak(s); Caribs
Italian language, 132, 133

Jahn, Friedrich Ludwig, 173
Jastal, Katarzyna, 173
Jespersen, Otto, on male/female language differences, 95, 206n179 (chap. 2)
"Jews," the, as linguistic alien (Wagner), 48–49
Jones, William Jervis, on women and Latin, 75–76
Juvenal, critical of women's speech (*Satires*), 75

Kant, Immanuel, 17–18, 53, 83; on race, 111
Katz, Marilyn, 199n40 (chap. 2)
Kilarski, Marcin, 119–20
Kirsteva, Julia, 10
Kittler, Friedrich, 54, 117, 129
Koerner, E. F. Konrad, 2–3
Kolb, Alexandra, 216n35 (chap. 4)
Kosch, Michelle, 43

Ladies' Book of Etiquette and Manual of Politeness, The (Hartley), 76
Lafitau, Joseph-François, on Ionian Greeks and Caribs (Herodotus), 69–70
Lakoff, Robin, 10, 182n46 (intro.)
Lamech: and adultery, 20–21; and word coinage, 20–21
Landes, Joan B., 25
langage des femmes, 67–68, 77–78. *See also* language(s); *Weibersprache*; women
language(s)
—articulation and differentiation as foundational, 172
—creation of, as men's work, 27, 29, 35, 60, 97, 113
—"crisis" of, 141–69
—defined, 28, 44, 53, 56, 172
—Europeans' compared with Carib, 68

—"feminine," 156–61; ironized, 161–64
—and form/matter binary (Humboldt), 55–57
—"genderised," 15, 171–72
—gender-neutral, opposed by far right, 170–75
—human body superior to (in dance), 150–51
—limits of, 141–42, 144
—masculinity of, 3, 12–13, 18, 58–59, 61, 174–75
—men's, 10, 64, 65, 71
—men's vs. women's primitivism, 70–73, 78, 79, 114–15, 162–63, 167, 206n179 (chap. 2)
—and metaphor, 60–61, 86–87
—mistrust of, 143
—and nature, 22, 57
—as organism, 107–9
—origins of (conjectural history), 107, 134, 213n731 (chap. 4); with Adam and Eve, 19–20; Adam Smith on, 24; with animals, 24; Benjamin on, 167; Brill on, 116; children and, 22–23, 186n37; Fichte on, 44–47; Formey on, 20; gender determination, 104; God and, 19–20, 185n22 (chap. 1); Grimm on, 59–60, 113–14; Herder on, 102–3; as human invention, 20–25, 35–38, 47–48, 52–53, 59, 61, 79; Humboldt on, 52–53, 172; Locke on, 20–21; as male (Fichte et al.), 12–13, 16–17, 18, 27–39, 47–48, 97, 113–16; Mandeville on, 22; Michaelis on, 22; Rousseau on, 17, 23–24; "savage" girls and, 25; Schiller on, 17; Schleicher on, 196n249 (chap. 1); unknowable, 60–61; Warburton on, 19–20
—pantomime as alternative to, 144
—and patriarchy, 28, 61, 169
—as representing different worldviews, 80
—"Romantic" conception of, 14, 119, 142
—secret (men's), 71–72
—"sexual" as superior/"scientific," 110, 111–12
—and sexual difference, 106–9, 113–19 (*see also* gender, grammatical)
—as speech/worldview, 54–55
—teaching of as national building, 117
—and thought as foundational, 172
—true lies in speech, 54–55
—vulgar, 94–95
—women's, 10, 13, 15, 57–58, 183n48 (intro.); as incoherent, 61; and body, 152, 166; dance as, 146–50; euphemism in, 91; as masculinist fantasy, 153; negative characterizations of, 73–84; as secret (Humboldt), 84–85; as silence, 70, 166; as solution to masculine "crisis," 142; universalized, 13, 72–73, 79–85
—women writing on, 14, 124–40
See also Arawak; Carib; gender, grammatical; "Island Carib"; males; women; *and individual languages and scholars*

—treatment of, 189n89 (chap. 1); measures
civilization level, 71, 72
—"types" of, 74–75
—violence against, 13, 64, 67, 69–70, 71–72,
86, 95
—"virtues" of, 32, 74–75; meekness in speech,
76–77; silence, 75, 143, 200–201n62
(chap. 2)
See also *Frauensprache*; males; marriage;
mother; patriarchy; *Weibersprache*

women's rights, Germans' conversation on, 87–88,
122–23. *See also* women
Woolf, Virginia, 124, 126
words: coinage of, 20–21, 24, 187n43 (chap. 1);
limitations of, 144, 151. *See also* language(s)

Yiddish, as inferior language (Wagner), 49

Zelle, Carsten, 143
Zipes, Jack, 119